D1696628

Omnidirectional Vision

The editors dedicate this book to their families.
They are grateful for their support throughout the course of this long project.

SCIENCES

Image, Field Director – Laure Blanc-Féraud

Sensors and Image Processing, Subject Head – Cédric Demonceaux

Omnidirectional Vision

From Theory to Applications

Coordinated by
Pascal Vasseur
Fabio Morbidi

WILEY

First published 2023 in Great Britain and the United States by ISTE Ltd and John Wiley & Sons, Inc.

Apart from any fair dealing for the purposes of research or private study, or criticism or review, as permitted under the Copyright, Designs and Patents Act 1988, this publication may only be reproduced, stored or transmitted, in any form or by any means, with the prior permission in writing of the publishers, or in the case of reprographic reproduction in accordance with the terms and licenses issued by the CLA. Enquiries concerning reproduction outside these terms should be sent to the publishers at the undermentioned address:

ISTE Ltd
27-37 St George's Road
London SW19 4EU
UK

www.iste.co.uk

John Wiley & Sons, Inc.
111 River Street
Hoboken, NJ 07030
USA

www.wiley.com

© ISTE Ltd 2023
The rights of Pascal Vasseur and Fabio Morbidi to be identified as the authors of this work have been asserted by them in accordance with the Copyright, Designs and Patents Act 1988.

Any opinions, findings, and conclusions or recommendations expressed in this material are those of the author(s), contributor(s) or editor(s) and do not necessarily reflect the views of ISTE Group.

Library of Congress Control Number: 2023938467

British Library Cataloguing-in-Publication Data
A CIP record for this book is available from the British Library
ISBN 978-1-78945-143-6

ERC code:
PE6 Computer Science and Informatics
 PE6_2 Computer systems, parallel/distributed systems, sensor networks, embedded systems, cyber-physical systems
 PE6_11 Machine learning, statistical data processing and applications using signal processing (e.g. speech, image, video)

Contents

Chapter 2. Models and Calibration Methods   39
Guillaume CARON

Chapter 3. Reconstruction of Environments   63
Maxime LHUILLIER

Conclusion and Perspectives . . . .
Fabio MORBIDI and Pascal VASSEUR

List of Authors

Index

Acknowledgments

This book would not have been possible without the support provided by our host institution, the University of Picardie Jules Verne, Amiens, France. Fabio Morbidi is indebted to his colleagues, E. Mouaddib and G. Caron, for the many refreshing discussions about omnidirectional vision over a beer.

The editors express their sincere thanks to the authors who have contributed to the six main chapters of this book, for the professionalism, reactivity, valuable comments and patience, which finally led to the publication of this collective work, after multiple delays due to the Covid-19 pandemic and our busy schedules. We learnt a lot from all of them, and this common journey strengthened our bonds of mutual esteem and friendship.

Finally, the editors wish to extend their heartfelt thanks to the team at ISTE Ltd for the smooth and efficient book production process, and to L. Blanc-Féraud (director of the "Image" field) and C. Demonceaux (subject head for "Sensors and Image Processing") for their guidance and encouragement.

Pascal VASSEUR and Fabio MORBIDI
Amiens, August 2023

List of Acronyms

AR: Augmented reality

BEV: Bird's eye view

CCD: Charge coupled device

CNN: Convolutional neural network

DNN: Deep neural network

DoF: Degrees of freedom

DoG: Difference of Gaussians

EKF: Extended Kalman filter

FoV: Field of view

GCM: General camera model

GEM: Generalized essential matrix

GPS: Global positioning system

GPU: Graphical processing unit

IMU: Inertial measurement unit

LM: Levenberg–Marquardt

MAP: Maximum a posteriori

MSCKF: Multi-state constraint Kalman filter

PnP: Perspective-n-point

RANSAC: RANdom SAmple Consensus

RGB-D: Red, green, blue – depth

RGBA: Red, green, blue, alpha

SfM: Structure-from-motion

SGM: Semi-global matching

SLAM: Simultaneous localization and mapping

SSD: Sum of squared differences

SURF: Speeded up robust features

SVD: Singular value decomposition

UAV: Unmanned aerial vehicle

VR: Virtual reality

WTA: Winner takes all

Preface

Fabio MORBIDI and Pascal VASSEUR

MIS Laboratory, University of Picardie Jules Verne, Amiens, France

P.1. Omnidirectional vision: a historical perspective

"Charge-coupled devices" (CCDs) were invented by W. Boyle and G.E. Smith at Bell Labs in 1969. They consist of a sensor that converts an incoming 2D light pattern into an electrical signal that, in turn, is transformed into an image. Although the CCDs could capture an image, they could not store it. Digital cameras, invented in 1975 at Eastman Kodak in Rochester, New York, by S.J. Sasson, fixed this problem. The first digital camera, equipped with a Fairchild Semiconductor's 100-by-100-pixel CCD, was able to display photos on a TV screen (Goodrich 2022).

An *omnidirectional camera* (also known as 360° camera) is a camera with a field of view (FoV) that covers approximately the entire sphere or at least a full circle in the horizontal plane (the adjective "omnidirectional" combines two words: "omni" and "directional". "Omni" comes from the Latin word "Omnis", meaning "all"). A conventional camera has an FoV that ranges from a few degrees to, at most, 180°: this means that it can capture, at most, light falling onto the camera focal point through a hemisphere. On the contrary, an ideal omnidirectional camera captures light from all directions falling onto the focal point, covering a full sphere. However, in practice, most omnidirectional cameras span only part of the full sphere and many cameras, which are dubbed omnidirectional, cover only approximately a hemisphere, or the full $360°$ along the equator of the sphere, the top and bottom hemispheres excluded (in this case, the term *panoramic camera* is preferred). If the full sphere is covered, the captured light beams do not exactly intersect in a single focal point, i.e. the system is non-central. Human vision is an example of a system with a wide FoV. In fact, humans have slightly over a $210°$ horizontal FoV (without eye movements) (Strasburger 2020), while some birds and insects have a complete or nearly complete $360°$ visual field. The vertical range of the visual field in humans is around $150°$. A large FoV has proven to be a very important asset in the preservation of certain species, and it likely plays a crucial role in the evolution of animal vision (Burkhardt 2005).

With the same ground being plowed many times by different researchers in the last decades, various camera designs have been proposed to capture $360°$ images: cameras with a single lens (fisheye), cameras with two lenses (dual or twin fisheye), cameras with more than two lenses (polydioptric), camera rigs, pan-tilt-zoom and cameras with rotating mechanisms, and catadioptric systems combining mirrors (cata-) and lenses (-dioptric). Some of these cameras capture $360°$ images in a single shot, while the others build an omnidirectional image by stitching together different regions of the FoV acquired over a prolonged period of time. Fisheye cameras (which use lens systems with very short focal lengths and strong refractive power), and catadioptric systems (which were first patented in 1970 (Rees 1970)), are in the first group. On the other hand, pan-tilt-zoom cameras and cameras with rotating mechanisms, belong to the second group.

Although the fundamental concepts have been around since at least the 1970s (Cao et al. 1986; Yagi and Kawato 1990; Ishiguro et al. 1992), *modern omnidirectional vision* dates back to the late 1990s: in fact, the seminal works on image formation and geometry by Nayar (1997); Baker and Nayar (1999); and Svoboda et al. (1998) marked the beginning of an independent field of investigation. Another milestone in the history of omnidirectional vision is the unifying theory for central panoramic systems developed by Geyer and Daniilidis (2000), and subsequently extended by Barreto (2006) and other researchers (Khomutenko et al. 2016; Usenko et al. 2018). This pioneering work has been the harbinger of a burgeoning array of papers in computer vision (image processing and descriptors adapted to spherical signals, calibration, epipolar and multi-view geometry, structure-from-motion, 3D reconstruction, etc.) and robotics (image-based localization, simultaneous localization and mapping (SLAM), visual servoing, etc.). Finally, the series of OMNIVIS workshops ("Omnidirectional Vision, Camera Networks and Non-Classical Cameras") held annually between 2000 and 2011, in conjunction with major computer vision conferences, contributed to shaping the community and bringing together researchers interested in non-conventional vision. This tradition continued with the OmniCV workshops ("Omnidirectional Computer Vision"), organized every year since 2020 (CVPR'23 marked the fourth edition).

Today, with the miniaturization of image sensors and optical components (lenses, prisms and mirrors), omnidirectional cameras have risen to prominence in consumer electronics (smartphone attachments, surveillance systems, perception systems in autonomous vehicles, etc.), and they have transformed our everyday lives and made them easier. Applications include mobile robotics, videoconferencing, art (panoramic photography), real estate (remote tours), vehicle parking assistance, virtual and augmented reality, tele-operated systems (for enhanced situational awareness), forensics, astronomy and entertainment. Several multinational electronics companies (Samsung, Ricoh, GoPro) have invested in the field of omnidirectional vision which has experienced a renaissance over the past ten years, and they are actively producing and supporting hardware. This fostered academic research and has contributed to the growth of the community.

P.2. Why this book?

While innumerable computer vision books have made their appearance in the last two decades, for example, books by Forsyth and Ponce (2011); Hartley and Zisserman (2004) and Ma et al. (2004) just to mention the most popular ones, relatively few books or monographs have been dedicated to omnidirectional vision. In fact, we are aware of only three research-oriented books (Benosman and Kang 2000; Sturm et al. 2011; Puig and Guerrero 2013), a survey paper (Ishiguro 2005), and two dedicated chapters in robotics textbooks, (Chapter 11.3 in Corke 2011; Chapter 4.2 in Siegwart et al. 2011). Several indicators suggest that the field of omnidirectional vision is now mature: it is then time to review the core principles (image formation, mathematical modeling, camera calibration, etc.), critically assess the key achievements and present some of the main applications, with an eye on the most recent trends and research directions.

Obviously, the field is too vast and dynamic to be fully covered in a single book. Therefore, a precise editorial choice has been made, and some "trendy topics" have been intentionally left out. A notable omission in coverage is the growing body of research on machine learning applied to omnidirectional vision (Ai et al. 2022), which is only briefly mentioned in Chapters 3 and 6. Moreover, we pass over the recent progress made in the field of graph image processing (Cheung et al. 2018). This book brings together the contributions of 10 renowned international scientists with multidisciplinary interests in image processing, computer vision, vehicle engineering and robotics. It is intended for a general audience: young beginners interested in discovering the field, professionals, instructors and experimented scientists in academia.

P.3. Organization of the book

This book consists of a preface, six chapters and a conclusion, and it is organized as follows:

– *Preface*: providing a brief history of omnidirectional vision, it defines the position and scope of the book, and presents its general structure.

– *Chapter 1* reviews basic geometric concepts relevant to omnidirectional vision. These include the image formation process, with a special focus on catadioptric cameras. A brief discussion on how camera models approximate the image formation process is also provided.

– *Chapter 2* presents the geometric models behind the formation of an omnidirectional image and critically assesses the different existing techniques for the estimation of the intrinsic parameters of an omnidirectional camera.

– *Chapter 3* describes different techniques for the reconstruction of 3D environments from images captured by static or moving omnidirectional cameras.

– *Chapter 4* is devoted to image processing, adapted to the spherical signals provided by catadioptric cameras.

– *Chapter 5* presents a special class of omnidirectional cameras, the so-called non-central vision sensors, and provides an overview of their main geometric properties and applications.

– *Chapter 6* deals with the application of omnidirectional cameras to robot localization and navigation.

– *Chapter 7* concludes the book. The main contributions are summarized and some prospects for future research are discussed.

> It is not knowledge, but the act of learning, not possession but the act of getting there, which grants the greatest enjoyment. When I have clarified and exhausted a subject, then I turn away from it, in order to go into darkness again. The never-satisfied man is so strange; if he has completed a structure, then it is not in order to dwell in it peacefully, but in order to begin another. I imagine the world conqueror must feel thus, who, after one kingdom is scarcely conquered, stretches out his arms for others.

Extract from a letter of Carl Friedrich GAUSS to Farkas BOLYAI, dated 2 September 1808.

June 2023

P.4. References

Ai, H., Cao, Z., Zhu, J., Bai, H., Chen, Y., Wang, L. (2022). Deep learning for omnidirectional vision: A survey and new perspectives [Online]. Available at: https://arxiv.org/abs/2205.10468.

Baker, S. and Nayar, S. (1999). A theory of single-viewpoint catadioptric image formation. *Int. J. Comput. Vision*, 35(2), 175–196.

Barreto, J. (2006). A unifying geometric representation for central projection systems. *Comput. Vis. Image Und.*, 103(3), 208–217.

Benosman, R. and Kang, S. (eds) (2000). *Panoramic Vision: Sensors, Theory, and Applications*. Springer-Verlag, New York.

Burkhardt Jr., R.W. (2005). *Patterns of Behavior: Konrad Lorenz, Niko Tinbergen, and the Founding of Ethology*. The University of Chicago Press.

Cao, Z., Oh, S., Hall, E. (1986). Dynamic omnidirectional vision for mobile robots. *J. Robotic Syst.*, 3(1), 5–17.

Cheung, G., Magli, E., Tanaka, Y., Ng, M. (2018). Graph spectral image processing. *Proc. IEEE*, 106(5), 907–930.

Corke, P. (2011). *Robotics, Vision and Control: Fundamental Algorithms in MATLAB*, volume 73. Springer-Verlag, Berlin/Heidelberg.

Forsyth, D. and Ponce, J. (2011). *Computer Vision: A Modern Approach*, 2nd edition. Pearson Education, Upper Saddle River.

Geyer, C. and Daniilidis, K. (2000). A unifying theory for central panoramic systems and practical implications. In *Proc. 6th Eur. Conf. Comput. Vis.* Springer, Berlin/Heidelberg.

Goodrich, J. (2022). The first digital camera was Kodak's biggest secret: The toaster-sized device displayed photos on a TV screen. *The Institute*, 60–61.

Hartley, R. and Zisserman, A. (2004). *Multiple View Geometry in Computer Vision*, 2nd edition. Cambridge University Press.

Ishiguro, H. (2005). Omnidirectional vision. In *Handbook of Pattern Recognition and Computer Vision*, Chen, C.H. and Wang, P.S.P. (eds). World Scientific, Singapore.

Ishiguro, H., Yamamoto, M., Tsuji, S. (1992). Omni-directional stereo. *IEEE Trans. Pattern Anal. Mach. Intell.*, 14(2), 257–262.

Khomutenko, B., Garcia, G., Martinet, P. (2016). An enhanced unified camera model. *IEEE Robot. Autonom. Lett.*, 1(1), 137–144.

Ma, Y., Soatto, S., Košecká, J., Sastry, S.S. (2004). *An Invitation to 3D Computer Vision: From Images to Geometric Models*. Springer-Verlag, New York.

Nayar, S. (1997). Catadioptric omnidirectional camera. In *Proc. IEEE Conf. Comput. Vis. Pattern Recognit.* IEEE, New York.

Puig, L. and Guerrero, J.J. (2013). *Omnidirectional Vision Systems: Calibration, Feature Extraction and 3D Information*. Springer-Verlag, London/Heidelberg.

Rees, D.W. (1970). Panoramic television viewing system. United States Patent, No. 3, 505, 465.

Siegwart, R., Nourbakhsh, I., Scaramuzza, D. (2011). *Introduction to Autonomous Mobile Robots*, 2nd edition. MIT Press, Cambridge, MA.

Strasburger, H. (2020). Seven myths on crowding and peripheral vision. *i-Perception*, 11(3), 1–46.

Sturm, P., Ramalingam, S., Tardif, J.-P., Gasparini, S., Barreto, J. (2011). Camera models and fundamental concepts used in geometric computer vision. *Foundations and Trends in Computer Graphics and Vision*, 6(1–2), 1–183.

Svoboda, T., Pajdla, T., Hlaváč, V. (1998). Epipolar geometry for panoramic cameras. In *Proc. Eur. Conf. Comp. Vis.* Springer, Berlin/Heidelberg.

Usenko, V., Demmel, N., Cremers, D. (2018). The double sphere camera model. In *Proc. IEEE Int. Conf. 3D Vision.* IEEE, New York.

Yagi, Y. and Kawato, S. (1990). Panorama scene analysis with conic projection. In *Proc. IEEE/RSJ Int. Conf. Intel. Robots Syst.*, volume 1. IEEE, New York.

1

Image Geometry

Peter STURM

Inria Grenoble Rhône-Alpes, Montbonnot-Saint-Martin, France

This chapter reviews basic geometrical concepts relevant for omnidirectional vision. These comprise the image formation process, with a special emphasis on catadioptric cameras, and a brief discussion on how camera models approximate the image formation process. The distinction between central (or single-viewpoint) and non-central cameras is explained, together with a discussion of the respective advantages. The chapter also reviews the basic building blocks of structure-from-motion, ranging from projection and back-projection to pose and ego-motion estimation. It is then shown that line images are particularly important when studying omnidirectional cameras for calibration as well as for image matching through the epipolar geometry. Dense matching is usually sped up by pre-processing images during a rectification process, which is explained with a particular emphasis on omnidirectional images.

The concepts discussed in this chapter are illustrated in short videos accessible on the Internet (creative commons license CC-BY-NC-SA). URLs are provided in the text.

For a color version of all figures in this chapter, see www.iste.co.uk/vasseur/omnidirectional.zip.

Omnidirectional Vision,
coordinated by Pascal VASSEUR and Fabio MORBIDI. © ISTE Ltd 2023.

1.1. Introduction

Geometry is important in various aspects of omnidirectional vision, from sensor and camera design and modeling, over image analysis, to structure from motion. The probably most fundamental issue concerns design and modeling: how to build image acquisition devices that have certain desired characteristics? Foremost among these is of course the desire to acquire images with a very wide field of view, be it panoramic, hemispheric, fully spherical, or somewhere in between[1]. The potential interests are clear – such a wide field of view allows us to visualize or analyze a scene more completely[2] and detect obstacles or objects to interact with, all around a robot. It also turns out that ego-motion estimation is generally more stable and accurate with a large field of view (Nelson and Aloimonos 1988). Besides such practical interests, there may also be others, such as esthetics.

Various technical solutions have been developed to acquire omnidirectional images. The first ones relied on rotating a regular or tailor-made camera about itself and "stitching together" images acquired during the rotation such as to form an image representing an extended field of view, an approach nowadays provided as a basic feature on most consumer-grade digital cameras. The obvious disadvantages are that image acquisition is not instantaneous, making it difficult to operate in dynamic contexts or to use it for omnidirectional video acquisition, and that it requires image processing that may not succeed for all types of scenes. Alternative solutions were thus developed, especially of two types. One consists of developing lens designs capable of delivering the sought after large fields of view, in particular fisheye objectives with fields of view nowadays even exceeding $180°$. The other, quite popular in robotics in the last decades, is to use mirrors to enhance the field of view of a camera, leading to systems baptized catadioptric cameras[3].

---

1 In the following, for brevity, we will mostly use the term omnidirectional when speaking of very large fields of view.

2 Among the first applications were the creation of images for panoramic viewing and the study of cloud formations from hemispheric images of the sky. Surveillance type applications are of course also a natural client.

3 Catadioptric meaning the combination of both refractive and reflective optical elements (lenses and mirrors).

Besides an extended field of view, various other design objectives have been pursued by scientists and engineers. Among the most important is the aim of achieving a single effective viewpoint or optical center – in the following we will also speak of *central cameras*. An example is given in Figure 1.1.

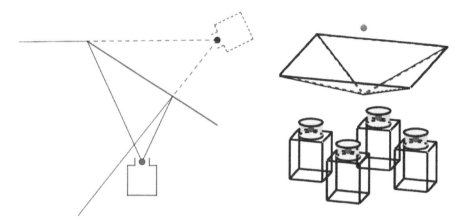

Figure 1.1. *Central catadioptric systems based on planar mirrors*

NOTES ON FIGURE 1.1.– *Left: A camera pointed at a planar mirror acquires the same image (besides issues such as loss of sharpness or color richness due to imperfect reflection) as would a virtual camera situated and oriented symmetrically on the other side of the mirror. Right: Several camera–mirror pairs are arranged such that the corresponding virtual cameras have the same optical centers. The arrangement on the right-hand side of the figure allows us to produce a panoramic image as if it were taken from a single effective viewpoint inside the pyramid of mirrors – it thus corresponds to a central camera (Iwerks 1964; Nalwa 1996). To get a complete panoramic field of view, an adequate number of camera–mirror pairs must be used, depending on the camera's individual fields of view.*

Why is this property interesting? Firstly, as noted by Baker and Nayar (1999), images acquired by central cameras allow it to synthesize perspectively correct images, that is, images as if acquired from the same viewpoint by a perspective camera. This property has been used to "navigate" in panoramic images through panning our virtual viewing direction and displaying a perspective image generated from the panorama according to the current viewing direction (Chen 1995). Obviously, this idea is applicable for any combination of real and virtual camera, provided they are both central and that the real camera offers a sufficient field of view to synthesize images for the virtual camera. Examples are shown in the accompanying videos 01_Perspective_Rendering.mp4.[4] and 02_Panoramic_Rendering.mp4[5], where it is illustrated how to synthesize perspective panoramic images from an image acquired with an omnidirectional catadioptric camera.

Another advantage of having a single effective viewpoint is that this opens the way to directly applying the rich toolkit of Structure-from-Motion (SfM) methods originally developed for perspective cameras. A final aspect mentioned here is that the usage of central cameras enables us to perform dense stereovision very efficiently, in a manner analogous to that originally developed for perspective images, through a generalized rectification process followed by "scan-line matching". All these aspects are further developed later in this chapter.

Other design objectives besides large field of view and single effective viewpoint concern for instance the types of distortion inevitably generated by omnidirectional cameras. Some cameras are built in order to conform to an equiangular (sometimes also called equidistant) distortion profile: this is the case if the angle spanned by the optical axis and the viewing direction associated with a pixel in the image is proportional to the pixel's distance from the distortion center in the image. Other distortion profiles may be interesting such as ones corresponding to area-preserving projections (Hicks and Perline 2002). The choice of distortion profile or other design feature may depend on the practical application of a camera, for instance, the requirement that certain parts of the scene appear in higher resolution than others. A general framework for specifying such properties and developing a dedicated mirror shape achieving these as best as possible was proposed by Swaminathan et al. (2004) and Hicks (2005).

Various other design objectives have of course been explored, such as on optical properties (ease of focusing, color fidelity, etc.) or the volume of a camera: to circumvent the classical disadvantage of catadioptric cameras concerning their bulk, "folded" catadioptric systems composed of several nested mirrors were proposed (Nayar and Peri 1999).

---

4 Available at: https://hal.inria.fr/hal-03563184.

5 Available at: https://hal.inria.fr/hal-03564938.

Let us close this discussion by mentioning that while a single effective viewpoint is an attractive property, as explained above, designing cameras that are non-central on purpose may be beneficial in other ways: this obviously gives more degrees of freedom to optimize certain properties and also non-central cameras may allow us to determine the absolute scale of ego-motion and 3D reconstruction, which is not possible with central cameras. Non-central cameras are further discussed in section 1.4.

In this chapter, we discuss geometrical aspects of omnidirectional cameras. When discussing "the geometry" of a certain type of camera, we usually refer to some model of the image formation process: how a camera generates an image of the scene it is pointed at. The actual image formation process carried out by a real camera is relatively complex; light emitted by objects in the scene travels through space and, if entering the camera aperture, travels across the camera's optical elements (lenses, mirrors) and finally hits the optical sensor, producing an electrical charge (for digital cameras) that is eventually converted into greylevels or colors for displaying an image to a human or for processing it by a computer. Most models of this process used in computer vision are simplified representations of it. In this chapter, we are concerned with purely geometric models, whose main basic operation is to determine, given the location of a point-like object in the scene, where the object's image will be observed on the sensor. All higher level geometric operations in computer vision, such as computing the motion of a camera from two or more images of a scene, are derived from this basic operation.

1.1.1. *Outline of this chapter*

In the next section, we briefly outline the general image formation process and explain in which ways the usual camera models, even the classical perspective or pinhole model, are a simplification thereof. Two fundamental concepts, projection and back-projection, are very briefly introduced in the subsequent section, followed by a short discussion on central and non-central cameras. When talking about 3D geometry for computer vision, a useful distinction is between what happens "outside" cameras from their "inner workings". By "outside", we essentially mean the geometric relations between the 3D scene and one or more cameras and more particularly, the question of how to represent and infer information on the relative positioning between cameras and/or entities in the scene. This is the subject of a section, where it is addressed, for conciseness, for the case of fully calibrated cameras. The subsequent two sections then consider the geometry of the inner workings of cameras and the epipolar geometry of a pair of cameras, with an attention to the subject of rectification for stereo matching.

This chapter does not contain equations, only geometry–algebraic formulations of some of the material covered can be found in other chapters of this book and another good starting point may be Sturm et al. (2011).

1.2. Image formation and point-wise approximation

The generation of a digital image of a scene, by means of a camera, is a complex process. It is ultimately created through photons that cause an electric response in the camera sensors, which are made of either CCD, CMOS or another technology. We may distinguish what happens inside a camera, to photons that enter its aperture, from what is going on outside: these photons result from a potentially infinitely complex game, being emitted, reflected, refracted, etc., from objects in the scene, where light is "bounced" repeatedly from one object to another. Not to speak of phenomena such as atmospheric or submarine diffraction caused by particles suspended in the air or water.

All of these aspects of the image formation process have been studied, in various degrees, in photogrammetry, computer vision and computer graphics, guided by the motivation to synthesize as realistic images as possible, to enable a meaningful analysis of imagery acquired in bad weather or otherwise "unusual" conditions such as under water, or even to "look around the corner" (Torralba and Freeman 2012), that is, to use an acquired image to infer something about an object hidden from the camera's field of view through its reflection or shadows produced on other objects.

Most works in computer vision rely on different levels of approximation of the image formation process. A first such approximation is to ignore the multiple bounces light undergoes in the scene: we implicitly assume that each point on the surface of an object in the scene emits/reflects light, and only photons reaching the camera on a direct path are considered. Sometimes, the "light emitted" by a point is modeled by a "color" associated with that point or, more generally, by a reflectance model which represents how light impinging on that point from a direct light source is reflected in different directions. In any such case, whenever the camera aperture is of finite extent (i.e. is not assumed to be a single point), the camera captures an entire volume of light emitted/reflected by any individual scene point. Unless the camera optics are perfect and the point is perfectly in focus, this set of light rays hits the camera sensor on a finite area, that is, the "image" of the scene point is not confined to an infinitesimal image point. This is of course one of the well-known causes of blur. Image formation models that mimic this have been proposed in computer vision and graphics, for example, through the definition of point spread functions[6].

A second level of approximation comes essentially down to ignoring blur and to assuming that light emitted by a single scene point manifests itself in a single point in the final image. Besides the above sketched cause for blur, this approximation also ignores the fact that in a real digital camera, light is captured through a finite set of photosensitive elements, each one capturing light within a finite area[7].

Most works tagged as "geometry" in computer vision use such an approximation: the image of a scene point is again a point. Camera models are then conceived to essentially allow one to compute, given the location and orientation of a camera and the location of a scene point, the location of the image point associated with that scene point. Such camera models are the basis for various tasks such as estimating the location and orientation of an object relative to a camera (pose estimation), estimating the motion of a camera just by analyzing different images taken during that motion, estimating a 3D model of the scene, etc. We will come back to these tasks in later sections. Before doing so, we first investigate, in the next section, camera models and how they represent the mapping of a scene point to an image point.

1.3. Projection and back-projection

The simplest and most widely used camera model in computer vision is the so-called pinhole model, performing a perspective projection[8]. It consists of two elements: the entire optics is represented by a single point, the so-called

---

6 Available at: https://en.wikipedia.org/wiki/Point_spread_function.

7 We also ignore aspects such as blooming across different photosensitive elements, black current and so on.

8 Even simpler are the affine approximations thereof, such as the orthographic model.

optical center, and the image sensor is represented by a mathematical plane, the image plane. The basic operation, the *projection* of a scene point to the image, is performed as follows: the first one creates the mathematical line that connects the optical center and the scene point and second one determines the point where that line intersects the image plane: the image point.

These simple geometrical operations can be expressed in a similarly simple algebraic manner (see, for instance, Sturm et al. 2011).

As explained in section 1.2, this model of course represents multiple approximations to the functioning of a real camera. Even so, it has been observed that it models "regular" cameras (e.g. consumer cameras) sufficiently well, that is, when being employed for tasks such as 3D reconstruction and motion estimation, the results are of acceptable accuracy. However, the approximation becomes insufficient when radial or other distortions become noticeable, for instance in wide field of view cameras, or whenever the maximum possible accuracy is sought after. This can be taken care of by extending the perspective camera model accordingly, as has been studied in photogrammetry for more than a century, resulting in more complex camera models (both geometrically and algebraically). However, even these classical extensions of the perspective model are not sufficient when considering omnidirectional cameras, we will consider these further below.

Let us now study an important concept, the reciprocal operation to projection, which we call *back-projection*. Back-projection starts from an image point and tries to answer the question where the original scene point could possibly be located. In general, unless additional information is available, the answer corresponds to a (half-) line. Back-projection for the perspective model is straightforward: we determine the line connecting the image point and the optical center, followed by "clipping" it to a half-line, for instance, at the optical center or at some minimal viewing distance in front of it. Both operations have a simple algebraic expression, much like for projection.

We now move toward the case of a catadioptric camera: a system composed of a perspective camera and a mirror into which this camera gazes[9]. Suppose that the entire geometry of the system is known: position of the image plane and the optical center, as well as the shape and position of the mirror. Let us first consider the case of a general mirror shape, that is, it is not constrained to be a surface of revolution or otherwise specific. Suppose that the mirror shape is expressed by a scalar function that takes 3D point coordinates as argument and returns the distance of the 3D point to the closest point on the mirror surface. Let us further suppose that we have a function that maps 3D points lying on the mirror surface to the associated mirror's tangent plane[10].

It is probably obvious that in the general case, neither projection nor back-projection have a closed-form solution. Let us now show that back-projection is in general simpler to perform than projection. Back-projection can be solved as follows. Start from a point in the perspective camera's image plane. We back-project from the perspective camera, as shown above, which results in a half-line determined by the image point and the optical center. To determine where the original 3D point could be located, we need to "back-track" the projection path. Here, we first have to determine where our half-line intersects the mirror. It should be clear that this can be done by some one-dimensional search procedure (search along the half-line). Once this intersection point is known, we retrieve the mirror's tangent plane at that point and reflect the half-line in it; the extension of the resulting line segment to a half-line constitutes the final result of the back-projection problem.

As for the *projection* problem, it turns out to be more complex: we need to determine a point on the mirror surface, such that the reflection in the associated tangent plane, of the line spanned by that point and the 3D scene point, is a line that contains the perspective camera's optical center. Once this is known, it merely remains to compute the final image point by intersecting this reflected line with the image plane. In general, the first step here, the determination of the appropriate point on the mirror surface comes down

9 Note that the camera gazing at the mirror may be non-perspective, for instance, a regular camera with radial distortions. But most works on catadioptric systems assume a perspective camera and for ease of exposition, we stick to this convention in the following.
10 The following observations also apply when using other representations of the mirror shape.

to a *two-dimensional* search. Hence, projection is more difficult than back-projection in this general catadioptric setting.

A few comments can be made. Firstly, if the mirror shape is a surface of revolution and if the perspective optical center lies on the axis of revolution, then projection can be reduced to a one-dimensional search, like for back-projection. Secondly, projection can actually lead to multiple solutions, at least if the mirror is not convex (relative to the "side" that is visible to the scene). This is shown in Figure 1.2. It shows a two-dimensional example where a point is actually projected to multiple image points. Two of these correspond to physically valid solutions, the others are only mathematical solutions but which cannot be realized in practice. Note however that in special cases, such as for central catadioptric cameras (see below), where closed-form algebraic formulations for projection exist, these generally deliver all mathematically valid solutions, not only the physically realizable ones. As for back-projection, there is only a single physically valid solution[11]. Additional *mathematical* solutions may still exist though, like for projection. A third comment is that, as already hinted on just above, certain mirror shapes enable a closed-form algebraic expression, especially for back-projection but possibly also for projection.

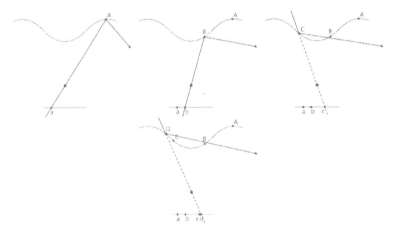

Figure 1.2. *A camera, composed of an optical center – in red – and an image plane, looks at a curved mirror (composed of three parabolic arcs). The goal is to find the possible image points of the scene point (in blue)*

11 Besides in cases where the mirror surface is allowed to be discontinuous.

NOTES ON FIGURE 1.2.– *The four possible image points are shown, in each case with the corresponding point on the mirror where the reflection takes place. Point a and b are physically correct image points, whereas c and d are spurious mathematical solutions (they lie on the infinite lines resulting from the reflection of the scene point in mirror points C and D). The video* 03_Three_Parabolae.mp4 *(see: https://hal.inria.fr/hal-03564950) shows an animation for this scenario.*

Figure 1.3 shows the same situation with an elliptical mirror. While there is only a single physically realizable image point in this case, there are in total four mathematical solutions. If the optical center of the camera is situated at one of the ellipse's foci, then the entire system becomes central (see next section) and also the four mathematical solutions collapse to two.

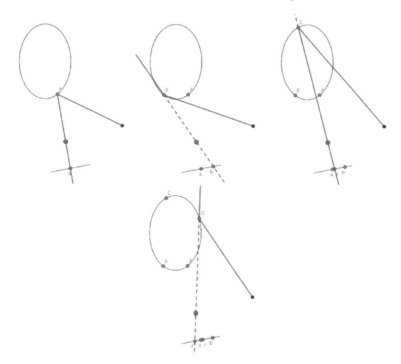

Figure 1.3. *Same situation as in Figure 1.2, but with an elliptical mirror. Only image point a is physically realizable, and the others are spurious mathematical solutions. See the video* 04_Ellipse_Projection.mp4 *(see: https://hal.inria.fr/hal-03564960) for an animation*

1.4. Central and non-central cameras

The notion of central camera has already been defined in section 1.1. Using the concept of back-projection described in section 1.3, we can define central cameras as those where all back-projection lines are incident with a point, the camera's (effective) optical center or viewpoint. Cameras not satisfying this condition are termed non-central. Special cases exist, such as cameras where all back-projection rays are incident with a line or line segment – in Ramalingam et al. (2006) these were called axial cameras.

In the following, we consider these notions in particular for catadioptric cameras. The first examples of central catadioptric systems are already given in section 1.1 (see Figure 1.1). The most trivial setup consists of a central camera looking into a planar mirror. And, as shown above, arranging several camera–mirror pairs appropriately leads to a central image acquisition device with a compound field of view that is fully panoramic.

The obvious main drawback of this design is the requirement of multiple cameras. How to obtain an omnidirectional field of view using a single camera and a single mirror, while simultaneously achieving a single effective viewpoint, was fully investigated in Drucker and Locke (1996) and Baker and Nayar (1999). The only such systems correspond to mirrors which are surfaces of revolution whose generatrix is a conic (that is symmetric in the axis of revolution). Furthermore, the central camera looking in the mirror must be positioned such that its optical center coincides with a focus point of the mirror[12]. The different possibilities for such systems are described in detail in Baker and Nayar (1999) – we can distinguish trivial and degenerate cases from practically useful ones as follows.

The first trivial/degenerate case was mentioned just above, a planar mirror. Its generatrix is a line orthogonal to the axis of revolution and the camera may be positioned anywhere. The next case corresponds to the generatrix being a pair of lines: the mirror surface is thus a circular cone[13]. A cone possesses a single focus, its apex. This is thus a degenerate case since a camera located at the cone's apex does not actually see the reflection of the scene in the mirror. Cone-shaped mirrors were still used to build catadioptric systems, but with

---

12 Note that the type of camera is irrelevant as long as it is central, that is, it may in principle be a perspective camera with distortions, a central fisheye, etc.

13 Only one of the two mathematical cone lobes is used.

cameras situated away from the apex, leading to a non-central system (see below).

The remaining three cases correspond to the generatrix being a parabola, ellipse or hyperbola, respectively. See Figure 1.4 for illustrations of the explanations given in the following.

The parabolic mirror has two foci: one "inside" the mirror and the other being the point at infinity of the mirror's axis of revolution. There are thus two possibilities to position the camera. Positioning it at the first focus point is rather impractical since it obstructs a large portion of the effective field of view and since the latter is not omnidirectional but rather narrow. It corresponds to the "bounding cylinder" of the mirror: the cylinder whose axis is that of the paraboloid and whose lateral extent is the same as that of the mirror. The second possibility is more interesting: the idea of positioning a camera at a viewpoint at infinity imposes that the camera realizes an affine projection, that is, where all viewing rays are parallel to one another and parallel to the mirror's axis. In practice, this can be achieved with an axis-aligned telecentric lens. For the reason explained above, the camera is positioned outside the parabolic mirror (see Figure 1.4; lower right part). The single effective viewpoint of this catadioptric system is the first focus point: indeed, when back-projecting image points from the affine camera and off the mirror, we obtain lines that are all incident with the focus point.

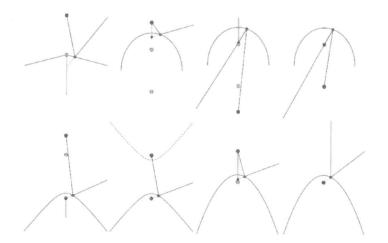

Figure 1.4. *Investigating central and non-central catadioptric cameras*

NOTES ON FIGURE 1.4.– *Each figure shows a case where a camera, represented by the optical center in red, is gazing at a mirror. The back-projection along a direction emanating from the optical center and through a point on the mirror (in green) is shown. The orange dots show the focus points of the mirror surface. The main aspect of interest here is where the back-projected rays intersect the mirror axis. In the upper left example (which corresponds to a planar section of a cone-shape mirror), the orange segment shows the locus of all possible such intersection points (when considering all possible back-projection directions). The fact that this locus is a segment and not a single point means that this catadioptric system is non-central. The three figures on the upper right correspond to an elliptical mirror. Only when the camera is positioned on one of the focus points does the locus collapse to a single point (identical with the other focus of the ellipse) and the system is a central one (right-most example). The two figures on the lower left show a hyperbolic mirror (the second one shows the two sheets of the mathematical hyperbola, whereas the actual mirror embodies a single one of these). Again, only if the camera is located at a focus point is the system central. The lower right concerns a parabolic mirror. In the second case, the system is central: the camera is "situated" at the parabola's focus point that is at infinity (thus not shown) – it must thus carry out an orthographic projection. Again, in this case, the above-mentioned locus collapses to a single point – the focus point that is "inside" the mirror.*

The second regular case consists of using an ellipsoidal mirror. Both foci lie inside the mirror shape. In practice, we would of course use a truncated ellipsoidal mirror surface, otherwise the camera would not see the scene outside the mirror. Still, this is, like the first parabolic case, of little practical interest due to a limited useful field of view. Note that the special case of a spherical mirror is actually fully degenerate. Here, the two foci collapse into a single one (the sphere center) and a camera situated there would see nothing else other than itself.

The last remaining case is that of a hyperboloidal mirror. To be precise, in practice we use a mirror corresponding to only one of the two sheets of a hyperboloidal surface. This has two foci, one inside and one outside the mirror's concave part. Positioning the camera on the second focus point is advantageous since the camera's reflection occupies a much smaller portion of the final image than in the reverse case.

Overall, there are in general thus two practically useful designs for single-mirror and single-camera central catadioptric systems based on paraboloid and hyperboloidal mirrors. Let us briefly examine the associated geometry of projection and back-projection. Although for a general camera–mirror system, projection involves (explicitly or implicitly) a search for a point on the mirror surface such that the scene point's reflection in the tangent plane at that point is directed toward the optical center (see section 1.3), the situation is considerably simplified for central catadioptric cameras. Namely, it suffices to compute the line joining the scene point and the mirror's second focus point to intersect that line with the mirror and to project that point into the perspective camera located at the other focus (or affine camera in the parabolic case)[14]. Back-projection works similarly: one firstly back-projects the image point to a camera ray, intersects this with the mirror and the computes the line joining the second focus and this intersection point[15].

While the geometric principle of projection and back-projection is the same for the parabolic and hyperbolic cases (and for the elliptic one too for that matter), it involves a focus point at infinity in the parabolic case and correspondingly, an affine camera instead of the perspective one used in the other cases. Note that a unified model that follows the same operations but that only handles finite entities, the so-called sphere model, has been proposed in Geyer and Daniilidis (2001) and Barreto and Araújo (2002). It simplifies the algebraic representation and geometric analysis of central catadioptric cameras.

Let us now briefly study non-central catadioptric systems, in particular systems using mirrors of the same type as above (surfaces of revolution with a conic as generatrix) but where the camera looking at them is not positioned at a focus point. A first observation is that if the camera's optical center lies on the mirror's axis of revolution, then the system is of the axial type: all back-projection rays are incident with a line, the axis of revolution, or rather, in general, a segment of that line. This is due to the following fact: since the

---

14 To be precise, there are two intersection points and thus two mathematical solutions for the projection.

15 Again, two mathematical solutions exist in general. In the parabolic case, one of the two solutions is always the same since the optical center – the point at infinity of the mirror axis – lies on the parabola. Hence, the mirror axis is always one of the two mathematical solutions for back-projection.

optical center lies on the mirror axis, any back-projection ray is coplanar with it, and since the mirror is a surface of revolution, the reflection of that ray in the mirror gives rise to a (half-) line that is again coplanar with the mirror axis. Hence, all final back-projection rays are incident with the mirror axis.

A few examples of this type are studied in Figure 1.4: the figure shows in particular the locus of effective viewpoints on the mirror axis, that is, the part of the mirror axis that is incident with back-projection rays of the system. In some cases, the locus is a line segment, whereas in others, it is a half-line or composed of two half-lines (not shown in the examples since corresponding to impractical settings). Note that in the case of a cone-shaped mirror for instance, an alternative choice to represent the viewpoint locus is possible instead of a line segment on the mirror axis: indeed, all back-projection rays are also incident with a particular circle centered in the mirror axis and "located" outside the cone (Baker and Nayar 1999).

If the optical center of the camera is not located on the mirror's axis, then the system is in general no longer axial and can be considered as fully non-central. Note that there are different "degrees of non-centrality": the most general one corresponds to the so-called oblique cameras (Pajdla 2002) where no two back-projection rays intersect one another.

Let us close this section by referring to section 1.1 where properties of central versus non-central cameras were discussed. Further differences are explained throughout the following sections, concerning structure from motion problems, epipolar geometry and dense stereo matching.

1.5. "Outer" geometry: calibrated cameras

In this chapter, we make the assumption that camera rays are straight (half-)lines, that is, light rays that form the camera image travel on straight paths before entering the aperture. More general situations occur, for instance, in airborne or satellite imagery, where atmospheric refraction may require us to loosen this assumption (an issue studied already decades ago in photogrammetry). Other similar examples concern scenes which are composed of multiple different media, such as cameras looking from the air into water, possibly through a curved glass window. The subjects handled in this section for straight camera rays have also been studied for such situations in the literature, giving rise to quite interesting geometric questions and findings (Maas 1995; Chari and Sturm 2009; Treibitz et al. 2012).

Given the scope of this chapter, we now return to the case of straight camera rays. We ask the question how to represent and infer information on the relative positioning between cameras and/or entities in the scene. This question has a very long history, which can be traced back at least to the Italian Renaissance and, for mathematical treatments, for at least about 300 years. In the following, we give a very concise overview of what we consider to be some of the main versions of this question.

1.5.1. *Given an image of a scene and a particular point in that image, where could the original point in the scene possibly be located?*

Since in this section we assume a fully calibrated camera, the immediate answer is trivial and it corresponds to the issue of back-projection already introduced: the scene point must be located somewhere on the (half)-line that can be computed/constructed from the calibration information. This is all we can say in the absence of further information. Additional information which may allow us to pin down the scene point's location more precisely is exploited in "bad weather imaging", where the amount of blur present around the image point, caused by fog in the scene, may give clues as to the depth of the scene point (Nayar and Narasimhan 1999)[16]. In this chapter, we do not consider such cases further and only use the "pure" geometrical information conveyed by (mathematical) points.

1.5.2. *Is it possible to precisely locate an object in 3D from a single image and if yes, what information is required to do so and how do we solve this problem mathematically?*

Here, an object is assimilated to a set of points. The main variant of this problem is nothing else than pose estimation or, as it is called in photogrammetry, resection: the required information is knowledge about the object's shape (i.e. the relative position of the object points). The minimal case where a general solution is possible is an object consisting of three points only. The "shape" of such an object is, for example, fully represented by the three pairwise distances between the points. The first work known to

---

16 Such approaches of course require further assumptions, such as on the homogeneity of the fog and the texture of the scene. Also, more generally, recent years have seen an increasing number of works where different cues are exploited for single-image 3D modeling through machine learning.

us that reports this problem and sketches a mathematical solution is by Lagrange (1773). He already showed that the problem can be reduced to finding the roots of a quartic polynomial and also sketches an iterative numerical procedure. While Lagrange did not explain in detail the construction of this quartic polynomial, he very likely had the complete solution. In 1841, a complete analytical solution was eventually provided by Grunert (1841).

How can we describe the pose estimation problem geometrically? One way of seeing it is as follows. Firstly, we back-project the three image points, giving three lines in 3D. Secondly, we try to displace the object (a triangle) in 3D, by translating and rotating it, such as to put it into a position where each of its three points comes to lie on the associated camera ray. This is something anyone can really try at home: arrange three spaghetti pieces any way you like and try to find out how to place a triangular object in the described way.

In general, there are up to eight different solutions to this problem. In the case of a central camera, as considered by Lagrange and Grunert, there are four pairs of mirror solutions (see Figure 1.5), whereas for non-central cameras, where the camera rays do not intersect in some common point, eight entirely different solutions may exist and where, incidentally, the analytical formulation comes down to finding the roots of an eighth-degree polynomial (Chen and Chang 2004; Nistér 2004a; Ramalingam et al. 2004).

Let us make a few complementary remarks on the pose estimation problem. Firstly, it can be stressed that the sketched minimal solutions are applicable to any type of central, respectively, non-central camera. For the central case for instance, it does not matter if the back-projection rays stem from a perspective camera, a fisheye, a catadioptric or any other central camera. Secondly, as is the case with many other minimal problems, the number of admissible solutions can in general be reduced when more than the minimum number of data (here, points) are available. More of this issue is discussed in the last part of this section. Finally, various special cases of the pose estimation problem have been covered in the literature, for example, the cases of a linear object or of an object consisting of more than three coplanar points, the case where the unknown pose has fewer degrees of freedom than six, etc.

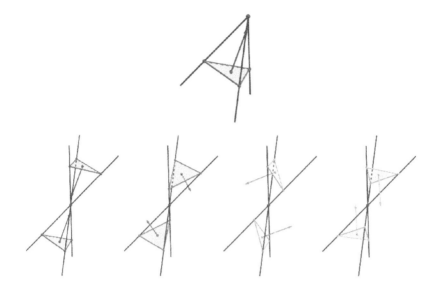

Figure 1.5. *Illustration of the three-point pose problem*

NOTES ON FIGURE 1.5.– *First: a central camera (only the optical center is shown, in red) gazing at three points. The camera rays are shown in blue. The three-point pose problem is to estimate the pose (position) of the three points, given only the camera rays (or, equivalently, the image points and the camera calibration) and knowledge of the pairwise distances between the original 3D points. Second: one solution is of course the original pose – a second mathematical solution is given by the mirrored pose (3D points symmetrically "behind" the camera). The remaining figures show three other pairs of solutions. Without further information (such as on the inclination of the normal vector of the plane spanned by the points), we cannot in general disambiguate the solutions.*

1.5.3. *Is it possible to estimate the motion of a camera just by taking images of an unknown scene?*

This important problem is called relative orientation in photogrammetry and motion or ego-motion estimation in computer vision. Maybe the first general solution is due to Kruppa who, in 1913, showed that five point correspondences between two images of a static scene are sufficient to solve this problem (Kruppa 1913). The minimal five-point problem is algebraically rather complex and only in 2004 was the first approach published, which provided only up to 10 theoretical solutions (previous approaches gave a higher number of solutions, some of which being spurious) (Nistér 2004b). Note that these approaches are designed for calibrated central cameras; for non-central ones, the minimal case is six point correspondences (Stewénius et al. 2005).

Geometrically, the motion estimation problem can be sketched as follows. Given the calibration information, the image points can first be back-projected. This gives two sets of 3D camera rays, one per image. It then suffices to align these two sets by rotating and translating them one against the other such that camera rays that belong to corresponding image points, intersect one another.

Unlike pose estimation, motion estimation is subject to a fundamental difference between central and non-central cameras. Let us first note that the problem has in principle six unknowns (three each for rotation and translation). However, if both cameras are central, then the translation can only be estimated up to scale, that is, the direction can be determined but not the extent of the translation as can be seen from Figure 1.6. If we have one possible relative motion, then moving the cameras along the baseline between the optical centers will leave the fact unchanged that corresponding camera rays intersect. This is the reason that five correspondences is the minimum case here. As for non-central cameras, this so-called scale ambiguity vanishes in general. This implies, on the one hand, that at least six correspondences are required and, on the other hand, camera motion is not only estimated up to scale, but precisely. However, this potentially strong advantage of non-central cameras has to be weighed with care: if the camera is only "slightly non-central", that is if the camera rays pass all through some small volume in space, then the accuracy of the estimated scale of translation may be poor. More on this issue is explained in section 3.6 of Sturm et al. (2011).

Note that, as for pose estimation, various special cases have been studied in the literature. The special case of a planar object was already solved by Schröter in 1880 (paragraph 45 of Schröter (1880)). Nistér (2004b) provides a method that directly solves the relative motion between three images, which gives, in many cases stabler results, and especially, works for planar scenes, in which case general two-view methods fail.

Let us finally remark that motion estimation is closely linked to epipolar geometry (see section 1.7). Especially for calibrated cameras, knowing the epipolar geometry is equivalent to knowing the relative motion between two images.

1.5.4. *Triangulation – reconstructing 3D points*

The last problem discussed here is that of reconstructing 3D points from image correspondences between calibrated cameras whose relative motion is known or has been estimated (to give, for instance, a 3D model such as the one shown in Figure 1.6). The minimal case corresponds obviously to two images. The first optimal method for this seemingly simple problem has been provided in Hartley and Sturm (1997), along with suboptimal methods for using correspondences in multiple images.

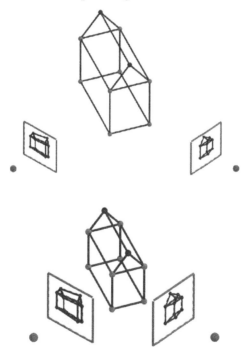

Figure 1.6. *Scale ambiguity of relative motion estimation with two central cameras: the relative motion can be estimated at best up to the scale of the translation between the cameras. The 3D reconstruction of the scene that can be computed using the estimated motion is affected by the same scale ambiguity. See an animation in the video* 05_Scale_Ambiguity.mp4 *(see: https://hal.inria.fr/hal-03564979)*

1.5.5. *Some remarks*

For the construction of complete and detailed 3D models, several to many input images as usually required. Many 3D modeling pipelines have been proposed in the literature. They are usually based on building blocks such as methods for the above or similar problems, complemented by approaches for extending them to multiple images.

In the remainder of this section, we will make a few remarks on the numerical methods used for solving these problems.

Above, we explained the minimum amount of information required to solve the examined problems. In practice, we usually have more data at our disposal, which opens the door to different ways of estimating the unknowns of the problem. We may distinguish the three main types of approaches.

Firstly, the so-called minimal methods, working exactly with the minimally required amount of information (three points for pose estimation, five for motion estimation, two views for triangulation and likewise for the mentioned special cases). These are usually expressed through (systems of) polynomial equations (see, for instance, Nistér 2004b; Hartley and Sturm 1997; Grunert 1841). Investigating minimal cases is of at least two-fold interest. On the one hand, this provides theoretical insights into studied problems, their complexity, the number of theoretically possible solutions, degenerate configurations where the problem is unsolvable, etc. On the other hand, minimal solvers are very relevant in practice in the case where the input may contain outliers, that is, wrong correspondences between points. As it is well known, such outliers, if not taken care of, may lead to completely erroneous numerical solutions to an estimation problem. One routine way of handling the outlier issue is to have recourse to random sampling approaches. The principle of the classical RANSAC (RANdom SAmple Consensus; Fischler and Bolles 1981) approach and its many variants is to repeatedly draw minimal random samples from the input data (e.g. triplets of points for pose estimation), to estimate hypothetical solutions to the problem from them and to check these against the remaining input data. With sufficiently many random samples, the probability of obtaining at least one sample with only inlier data is high, and in this case the consistency check with the other input data will signal a good consensus since, roughly speaking, most inliers will "signal agreement" with the hypothetical solution provided by this sample. The fact that a sample may provide multiple hypothetical solutions (such as up to four in the case of pose estimation) is handled trivially by checking the consistency of each of these solutions with the other input data. RANSAC is just one approach for robust estimation. For other possibilities, see, for instance, Huber (1981) and Triggs et al. (1999).

A second type of approach consists of trying to estimate a solution by using more than the minimum required data at once through solving a system of linear equations. For instance, how to approach pose estimation in this manner was explained in Quan and Lan (1999), and for the triangulation problem with an arbitrary number of input images, see Hartley and Sturm (1997). Such approaches have been quite popular in computer vision as they are computationally rather simple. A drawback however is that for most problems, the linear solution is suboptimal, as explained below.

The third main type of approach aims at a rigorous handling of uncertainty in the data. Ideally, we try to estimate the uncertainty of all input data and propagate it throughout the estimation process. For instance, when extracting interest points in images, we may estimate, besides the point positions, covariance matrices for the latter. The computation of problem variables (camera pose, relative motion, etc.) can then be formulated as a maximum a posteriori (MAP) estimation problem, where the likelihood is evaluated against these covariance matrices on the input data (or any other representation of their uncertainty). For most geometric computer vision problems, this then comes down to solving a nonlinear optimization problem. The general problem where the location of 3D scene points as well as camera poses are to be estimated is traditionally called bundle adjustment (see Triggs et al. (1999) for a general overview). As mentioned above, the objective function for these MAP formulations is nonlinear for most problems. Conversely, the linear formulations mentioned above are equivalent to solving maximum likelihood problems whose objective functions do not assess the quality of the estimates against the uncertainty in the input data, which is one explanation for their suboptimal performance.

Besides the above three types of approach, let us mention that for some problems, global optimization methods have been devised (see, for instance Hartley and Kahl (2009)).

1.6. "Inner" geometry: images of lines

One of the most interesting and important aspects of the geometry of omnidirectional cameras is how lines appear in the image. Although the image of a point is obviously a point itself under the assumptions stated earlier (or a set of points), the omnidirectional image of a line is in general not a line (see Figure 1.7).

The study of line images is interesting in several respects. Firstly, they tell much about the calibration of an omnidirectional camera and using lines is an attractive approach for calibration. Secondly, the geometry of line images is directly related to epipolar geometry and thus to (dense) binocular stereovision. In the following, we explain the link between line images and camera calibration; the issue of epipolar geometry is handled in the next section.

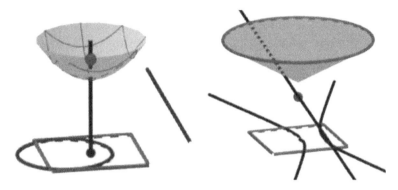

Figure 1.7. *Two examples of a catadioptric line image. Left: with a central catadioptric camera, line images are circles. Right: this system based on a cone-shaped mirror is non-central: the line image is not a conic as with all central catadioptric systems, but a higher degree curve (note for instance the small cusp in the curve on the left)*

Let us consider the example of central catadioptric cameras. It is well known that for these, the image of a line is a conic (Barreto and Araújo 2005). To be precise, this is only true in general if the camera looking at the mirror is a perspective one: otherwise, if the camera is subject to radial or other distortions, the line image is no longer a conic, but a conic "distorted" accordingly. But let us adopt the assumption of a perspective camera and thus the fact that line images are conics.

Let us provide a simple counting argument to explain why line images contain information on the camera calibration. It is well known that the set of lines in 3D space has four degrees of freedom. Now, consider a central camera and the image of some 3D line L. It is clear that any other line that lies in the plane spanned by L and the camera's optical center gives rise to the same line image as L (with the exception of lines going through the optical center, which constitute a singular case). Such lines represent a set with two degrees of freedom. This suggests that the set of images of all 3D lines in a given camera only has two degrees of freedom $(4 - 2)$. If the camera is perspective (and looking directly at the scene, i.e. not through a mirror or other device), then line images are lines themselves. The set of lines in the image plane has two degrees of freedom. This illustrates that for perspective cameras, images of lines do not convey any information on the camera's calibration, since no excess information is contained in them. In other words, for any line in the image plane, there exist 3D lines that are projected onto that image line. Hence, knowing lines images is not providing useful information for calibrating perspective cameras.

The situation is different for central catadioptric cameras. As already mentioned, line images are conics. The set of all conics has five degrees of freedom but as already explained, the set of line images only has two of them. In other words, not every conic in the image plane is a possible line image. Hence, knowing line images does tell us something about the camera's calibration. Calibration methods for catadioptric cameras that exploit this fact have been proposed in Geyer and Daniilidis (2002) and Barreto and Araújo (2005).

The situation is analogous for other types of omnidirectional cameras and for non-perspective cameras in general. For instance, using lines to calibrate non-perspective distortions in regular cameras is a classical approach called plumbline calibration (Brown 1971). As for perspective cameras with radial or other distortions, plumbline calibration allows us in general to estimate all calibration information about the camera, up to the perspective part (i.e. focal length, principal point, aspect ratio and coordinate axis shearing). As for omnidirectional cameras, line images may allow for the recovery of the entire calibration information. A formal study of the conditions in which full calibration is possible from line images does not exist to the knowledge of the author, but would be an interesting endeavor.

So far, we have only discussed the case of central (catadioptric) cameras. For non-central ones, the situation differs in that the images of any two 3D lines may differ from one another, hence that the set of line images now could have up to four degrees of freedom and not only two. Line images may still convey useful information for calibration. A general study of plumbline calibration for non-central cameras is another topic worthy of exploration.

1.7. Epipolar geometry

Epipolar geometry is a fundamental topic in computer vision and photogrammetry[17]. It essentially arises from the following question: given two images of a scene and a point in one of these, where could the corresponding point possibly be located in the other image? This question is one ingredient of the image matching problem, which is a prerequisite to most structure-from-motion tasks and dense stereovision. Studies of epipolar geometry usually concern at least two aspects of the question. Firstly, what can we say about the nature of the locus of corresponding points? For perspective cameras for example, the locus is a straight line, whereas for central catadioptric cameras, it is a conic. Secondly, what is needed to compute that locus? How can we compute it from information on the calibration and the relative pose of the two cameras? Or how can we compute it from a (potentially small) set of already available image matches? In the following, we only address the first aspect, followed by a discussion of implications for dense binocular stereo and rectification.

1.7.1. *Nature of epipolar geometry*

We address here the first question: given a point in one image, what can we say about the nature of the locus of corresponding points in the other image? The immediate answer directly follows from the considerations on line images exposed in the previous section and on back-projection (see section 1.3). Namely, we can simply divide the problem into two parts. Firstly, we ask the question where the original 3D point in the scene could possibly be located – this is simply the back-projection problem. As already seen, this gives rise to a (half-) line in 3D: without any additional

---

17 The oldest work known to us is Terrero (1862); the actual term epipolar geometry was introduced much later.

information, the original point may be located anywhere on it. Secondly, the locus of corresponding points in the second image is then nothing else than the image of that (half-) line.

Let us call that line image the *epipolar curve* of the first image point, in analogy to *epipolar lines* of perspective images. The first answer to our question is as follows: the corresponding point in the other image must lie on an epipolar curve and the nature of that curve is nothing else than the nature of line images of the second camera, that is, conics for central catadioptric cameras, lines for perspective cameras (see Figure 1.8) and so on.

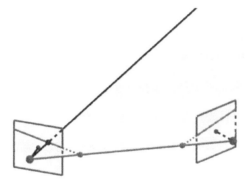

Figure 1.8. *Epipolar geometry of two perspective cameras.*

NOTES ON FIGURE 1.8.– *The image point in blue gives rise to a back-projection ray, whose image in the second camera is an epipolar line. The epipolar plane is the plane spanned by the back-projection ray and the optical center of the second camera – the epipolar line is the intersection of that plane with the image plane. The epipole of each camera (green point) is where the baseline between the cameras intersects the image plane (here, the epipoles are situated outside the image area, but this is irrelevant for most considerations). All possible epipolar lines are incident with the epipole, that is, they form a line pencil, with a single degree of freedom.*

It is important to realize that the nature of epipolar curves in one image depends only on the characteristics of the camera acquiring that image and not on the those of the other camera. In other words, whatever type the other camera is, it does not influence the nature of these epipolar curves, even if the other camera is non-central or otherwise different from the considered one.

The discussion so far concerns the nature of individual epipolar curves. To fully study epipolar geometry, we should also examine the characteristics of the set of all possible epipolar curves. Here, the type of the other camera does play its role. Let us explain this: whichever type the other camera is, back-projection of individual image points always gives rise to lines in 3D, hence the observations made above on the nature of epipolar curves. However, the composition of the set of all possible back-projection lines obviously depends on the type of camera. For a central camera, all back-projection lines go through the optical center, whereas for non-central cameras, this is by definition not the case.

These observations lead to several special cases depending on the types of the two cameras. A few of them are examined in the following. Firstly, the classical case of two perspective cameras (see the accompanying video 06_Epipolar_Geometry_Standard_Rectification.mp4[18]). Let us look at the set of back-projection lines of the first camera: this is a (subset of a) bundle of lines, all going through the optical center. The bundle of lines has two degrees of freedom. However, the set of the images of these lines in the other camera only has one degree of freedom. To see this, let us consider any one line in the bundle and the plane spanned by that line and the optical center of the second camera (a so-called epipolar plane). That plane contains a one-dimensional subset of lines in the bundle. All of them get projected to the same line image in the second camera. Overall, the lines in the bundle may be "partitioned" in a one-dimensional family of sets of lines lying in one epipolar plane each and with each set having the same line image. Thus, the set of epipolar lines in the second camera is one-dimensional only. Furthermore, as it is well known, all epipolar lines go through one particular point, the epipole, which is nothing else than the perspective projection of the first optical center in the second camera.

What if we replace the first camera by any non-perspective but still central camera? Well, as far as this discussion is concerned, nothing changes: epipolar curves in the second camera are still the same set of epipolar lines, all incident with the same epipole.

How about the epipolar geometry if the second camera is for example a central catadioptric one and the first one any central camera? Then, as

---

18 Available at: https://hal.inria.fr/hal-03564990.

already discussed, the epipolar curves in the second camera form a two degree-of-freedom set of conics. What about the epipole? In this case, since central catadioptric cameras map each 3D point to two image points, there are a pair of epipoles (Figure 1.9) and all epipolar conics are incident with both of them. As it is now clear, this observation is independent of the precise type of the first camera, as long as it is central.

Figure 1.9. *Epipolar geometry of two para-catadioptric cameras. Left: the two epipoles in each image are the two (mathematical) image points of the other camera's effective optical center (in red). Right: epipolar curve (circle in this case) in the second camera, being the image of the back-projection ray associated with the shown image point in the first camera. More on this is shown in the video* 07_Epipolar_Geometry_Para.mp4 *(see: https://hal.inria.fr/hal-03564997)*

Similar findings hold for other types of central cameras, for example, fisheyes: the nature of the epipolar curves is that of line images and the set of epipolar curves are a particular subset of admissible line images, all being incident with the epipole (or epipoles in case the camera maps 3D points to more than one image point each).

Let us now consider the case where the second camera is perspective but the first one non-central. A first example is that of a catadioptric camera with a cone-shaped mirror such that the camera looking at the mirror is situated on the mirror's axis. As shown above, this is a non-central device. All back-projection lines are incident with the mirror's axis and more specifically, with the line segment representing the caustic of the system (see Figure 1.4). Now, if the perspective camera is also positioned on the mirror's axis, then the back-projection lines are projected to a one-dimensional set of image lines, all incident with the point that corresponds to the "image" of the mirror axis, a point that can be considered as epipole. The situation differs as soon as the perspective camera moves away from the mirror's axis. Now the back-projection lines are mapped to a two-dimensional subset of the lines in the perspective image plane. To be precise, the epipolar lines are all lines in the perspective image plane, which are incident with the line segment that is the image of the catadioptric system's caustic. There thus no longer exists a distinct epipole; at best, we may consider the said line segment as playing a somewhat analogous though weaker role as that of a "regular" epipole.

Figure 1.10 shows examples of the epipolar geometry of two non-central cameras, catadioptric cameras with a cone-shaped mirror[19]. In the first case, when the two cameras are arranged on top of each other, the epipolar geometry is similar to that of perspective cameras, that is, epipolar curves are straight lines and there exist epipoles. This is no longer true in the second case.

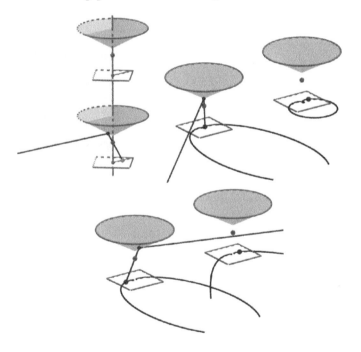

Figure 1.10. *Epipolar geometry of two non-central cameras; here, catadioptric systems with cone-shaped mirrors*

19 See also videos 08_Cones_Aligned_Epipolar_Geometry.mp4 (see: https://hal.inria.fr/ hal-03565001) and 09_Cone_Projection_And_Noncentral_Epipolar_Geometry.mp4 (see: https://hal.inria.fr/hal-03565016).

NOTES ON FIGURE 1.10.– *Left: if the two are placed on top of each other, the epipolar geometry is similar to that of perspective cameras. For ease of comprehension, only one half of the epipolar plane is shown. Middle and right: now the two systems are placed side by side. The figures show the epipolar curve in the first camera, associated with the image point in the second camera. They then consider a point in the first image, located on this epipolar curve, in two different positions. The epipolar curves in the second camera, associated with these two image points, differ from one another. This indicates that epipolar curves are no longer in a one-to-one correspondence here, unlike with central cameras.*

An interesting question is if such a "weak" epipolar geometry is still useful for image matching. There are two answers to this. Firstly, if the task at hand is to find the corresponding point of one individual image point in the catadioptric image, then the answer is positive: the corresponding point must lie on the epipolar curve associated with the point in the first, catadioptric, image and it does not matter that this epipolar curve is part of a weak epipolar geometry in the sense described. If however the task to be carried out is dense stereo matching, then the answer turns out to be negative in general. To examine this, we will first return to the case of two perspective images and recap how dense stereo matching can be, and usually is, carried out in that case.

1.7.2. *Dense stereo matching and rectification*

The standard pipeline for dense stereo matching is to first rectify the image pair and then to perform so-called scanline matching. Let us describe the rationale behind and workings of this procedure. The first essential aspect is that epipolar lines in two perspective images come in pairs: consider some point in the first image and the associated epipolar line in the second image. It turns out that for any point on that line, the associated epipolar line in the first image is always the same line (the line spanned by the first point and the epipole in the first image). There is thus a one-to-one correspondence: to each epipolar line in one image corresponds exactly one epipolar line in the other image. So, to find the matches to points on one epipolar line, it suffices to search among the points on the associated epipolar line in the other image. This means that the "complexity" of dense stereo matching is lower than it might appear at first: instead of doing a one-dimensional search each for a two-dimensional set of image points, it is possible to partition the problem into a one-dimensional set of problems, each of which consists of matching all the points on corresponding epipolar lines to one another in a single process.

The main advantage of the geometry at hand comes into play however through the way we can implement this last problem efficiently – this is not so much of a geometric issue but more of a computational one, in that it relies essentially on organizing computations and memory access efficiently. Let us first consider the initial situation where the goal is to find the matching point to a point in the first image and where the epipolar line is not axis-parallel. Point matching in general relies on comparing greylevels in windows around image points and computing some similarity measure from them. To find the matching point along the epipolar line, the basic procedure is thus to sample the latter by considering successive points on it, computing the similarity measure for each of those and then choosing the point that maximizes the similarity measure.

A first main source of the computational cost of this is memory access: each time a new point on the epipolar line is considered, we must access the greylevels of the pixels around it and this requires costly non-local memory accesses. A solution to circumvent this is to carry out a preprocessing where greylevels of each image are rearranged in memory in the following way: they are stored in a matrix such that greylevels of points on the same epipolar line appear in the same row. Further, neighborship relations should be preserved, meaning that greylevels of neighboring points (within an epipolar line as well as across neighboring lines) appear next to each other in that matrix. With this way of organizing the image information, the cost of memory access for the actual matching stage is much reduced: successively exploring all points on an epipolar line can be done through local memory accesses (simply put, by exploring sequentially arranged data in memory)[20].

This description of how to rearrange the information contained in the images for efficient stereo matching is actually just an alternative way of

---

20 A second important way of reducing computational cost of dense stereo matching is through avoiding redundant computations. For instance, when computing the similarity measure for a pair of points and for the pair consisting of the two points' right neighbors, some if not most of the computations done in the two cases will be identical. Through appropriate bookkeeping, the overall computational cost can be easily decreased dramatically.

explaining the rationale of the classical process of stereo rectification, as illustrated in Figure 1.11.

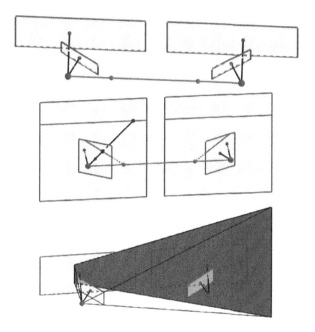

Figure 1.11. *Classical stereo rectification for perspective images. Upper left: one defines two coplanar virtual images planes that are parallel to the baseline. Upper right: one "projects" epipolar lines onto these virtual image planes. There, they will be axis-aligned and occupy identical image rows, allowing for efficient implementation of dense binocular stereo. Bottom: the required area for the first rectified image (camera slightly more panned inside than in the other figures). See also the video* 06_Epipolar_Geometry_Standard_Rectification.mp4 *(see:* https://hal.inria.fr/hal-03564990)

A significant drawback of this method for rectification[21] is that it cannot be applied when any of the epipoles lies inside the image area and that it becomes impractical when epipoles get close to the image area. The reason is that the required image area for the rectified images becomes prohibitively if not infinitely large (see Figure 1.11). This situation is of course frequent with omnidirectional images: with fields of view extending $180°$ and when epipoles

---

21 And more general ones, such as reasoning on homographies that map epipoles to points at infinity in the image (Hartley and Kahl 2009).

exist (e.g. with central cameras), at least one of them will always lie within the image area.

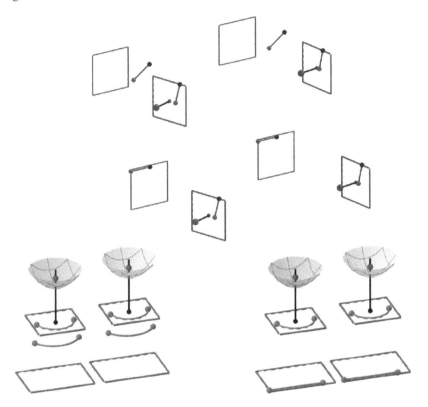

Figure 1.12. *Generic stereo rectification. Upper left: epipolar segments are "cut out" from original images and "pasted" into the rectified ones. Upper right: after the pasting. Bottom: the same has been used for para-catadioptric images. See videos* 10_Epipolar_Geometry_General_Rectification.mp4 *(see: https://hal.inria.fr/hal-03565020) and* 07_Epipolar_Geometry_Para.mp4 *(see: https://hal.inria.fr/hal-03564997) for animations and more explanations*

Alternative methods for rectification circumvent this problem, in an altogether simple manner (Pollefeys et al. 1999; Geyer and Daniilidis 2003). Their rationale is essentially as follows. Instead of rectifying epipolar lines through a perspective projection onto a virtual image plane, they simply "cut out" the line segments occupied by epipolar lines in the image area, and "paste" them into rows of the rectified images to be generated, as illustrated in Figure 1.12. All that needs to be done is to paste corresponding segments in the two images to corresponding rows in the rectified images, and to take care of the pixels that neighbor each other in the original images, which end up as neighbors in the rectified ones.

This idea, first developed for perspective images, can be readily extended to omnidirectional ones. The only essential additional step is that subsets of epipolar curves instead of line segments are concerned; it suffices to "unroll" these in order to paste them into the rectified images, as illustrated in Figure 1.12. With this simple procedure, we thus generate rectified image pairs that can in principle be processed using standard dense binocular stereo methods. Naturally, when exploiting the stereo matching results in order to generate, for instance, a dense depth map, we may need to map the matches obtained for rectified images, back to the original images, so that calibration information for these can be used for 3D reconstruction.

Let us conclude this section by reminding us that the application of this rectification principle requires the epipolar curves to be in one-to-one correspondence, in the sense explained above. This is always the case when the cameras are central and sometimes even when they are non-central, such as with cone-based catadioptric cameras on top of each other, as illustrated above. With non-central cameras in the general position however, this property breaks down. While epipolar curves continue to exist, they are in general no longer in one-to-one correspondence across the images. A general study on all possible stereo images is provided in Seitz and Kim (2002).

1.8. Conclusion

This chapter has revisited some fundamental geometrical aspects of camera models, with an emphasis on omnidirectional ones. Many topics were not discussed, such as multi-view matching relations (for more than two images), calibration or self-calibration. These, as well as algebraic expressions for the concepts presented, can be readily found elsewhere, for instance, in other chapters of this book.

1.9. Acknowledgments

The figures for this chapter were generated using the GeoGebra software (https://www.geogebra.org/). I wish to thank its developers for creating this software and making it freely available.

1.10. References

Baker, S. and Nayar, S. (1999). A theory of single-viewpoint catadioptric image formation. *Int. J. Comput. Vision*, 35(2), 175–196.

Barreto, J.P. and Araújo, H. (2002). Geometric properties of central catadioptric line images. In *Proc. 7th Eur. Conf. Comp. Vis.*, 4, 237–251. Association for Computing Machinery, New York.

Barreto, J.P. and Araújo, H. (2005). Geometric properties of central catadioptric line images and their application in calibration. *IEEE Trans. on Pattern Analysis and Machine Intelligence*, 27(8), 1327–1333.

Brown, D. (1971). Close-range camera calibration. *Photogramm. Eng.*, 37(8), 855–866.

Chari, V. and Sturm, P. (2009). Multi-view geometry of the refractive plane. In *Proc. British Mach. Vis. Conf.* British Machine Vision Association, Durham.

Chen, S. (1995). QuickTime VR – An image-based approach to virtual environment navigation. In *Proc. SIGGRAPH*. ACM SIGGRAPH, New York.

Chen, C.-S. and Chang, W.-Y. (2004). On pose recovery for generalized visual sensors. *IEEE Trans. Pattern Anal. Mach. Intell.*, 26(7), 848–861.

Drucker, D. and Locke, P. (1996). A natural classification of curves and surfaces with reflection properties. *Math. Mag.*, 69(4), 249–256.

Fischler, M. and Bolles, R. (1981). Random sample consensus: A paradigm for model fitting with applications to image analysis and automated cartography. *Graphics and Image Processing*, 24(6), 381–395.

Geyer, C. and Daniilidis, K. (2001). Catadioptric projective geometry. *Int. J. Comput. Vision*, 45(3), 223–243.

Geyer, C. and Daniilidis, K. (2002). Paracatadioptric camera calibration. *IEEE Trans. Pattern Anal. Mach. Intell.*, 24(5), 687–695.

Geyer, C. and Daniilidis, K. (2003). Conformal rectification of omnidirectional stereo pairs. *Proc. IEEE Conf. Comp. Vis. Pattern Recogn.*, 7, 73.

Grunert, J. (1841). Das pothenot'sche Problem in erweiterter Gestalt; nebst Bemerkungen über seine Anwendung in der Geodäsie. *Archiv der Mathematik und Physik*, 1, 238–248.

Hartley, R. and Kahl, F. (2009). Global optimization through rotation space search. *Int. J. Comput. Vision*, 82(1), 64–79.

Hartley, R. and Sturm, P. (1997). Triangulation. *Comput. Vis. Image Und.*, 68(2), 146–157.

Hicks, R. (2005). Designing a mirror to realize a given projection. *J. Opt. Soc. Am. A*, 22(2), 323–330.

Hicks, R. and Perline, R. (2002). Equi-areal catadioptric sensors. In *Proc. Workshop on Omnidirectional Vision*. IEEE, New York.

Huber, P. (1981). *Robust Statistics*. John Wiley & Sons, Hoboken.

Iwerks, U. (1964). Panoramic motion picture camera arrangement. U.S. patent 3,118,340'.

Kruppa, F. (1913). Zur Ermittlung eines Objektes aus zwei Perspektiven mit innerer Orientierung. *Sitzungsberichte der mathematisch-naturwissenschaftlichen Klasse der kaiserlichen Akademie der Wissenschaften, Abteilung II a*, 122, 1939–1948.

Lagrange, J.-L. (1773). Solutions analytiques de quelques problèmes sur les pyramides triangulaires. In *Nouveaux mémoires de l'Académie royale des sciences et belles-lettres*, Serret, J-.A. (ed.). Reprinted in 1869 in the 3rd volume of the *Œuvres de Lagrange*. Gauthier-Villars.

Maas, H.-G. (1995). New developments in multimedia photogrammetry. In *Optical 3-D Measurement Techniques III*, Grün, A. and Kahmen, H. (eds). Wichmann Verlag, Karlsruhe.

Nalwa, V. (1996). A true omnidirectional viewer. Technical report BL0115500-960115-01, AT&T Bell Laboratories, Murray Hill.

Nayar, S. and Narasimhan, S. (1999). Vision in bad weather. In *Proc. IEEE Int. Conf. Comp. Vis.* IEEE, New York.

Nayar, S. and Peri, V. (1999). Folded catadioptric cameras. In *Proc. IEEE Conf. Comp. Vis. Pattern Recogn.* IEEE, New York.

Nelson, R. and Aloimonos, J. (1988). Finding motion parameters from spherical motion fields (or the advantages of having eyes in the back of your head). *Biol. Cybern.*, 58(4), 261–273.

Nistér, D. (2004a). A minimal solution to the generalized 3-point pose problem. In *Proc. IEEE Conf. Comp. Vis. Pattern Recogn.* IEEE, New York.

Nistér, D. (2004b). An efficient solution to the five-point relative pose problem. *IEEE Trans. Pattern Anal. Mach. Intell.*, 26(6), 756–770.

Pajdla, T. (2002). Stereo with oblique cameras. *Int. J. Comput. Vision*, 47(1–3), 161–170.

Pollefeys, M., Koch, R., Gool, L.V. (1999). A simple and efficient rectification method for general motion. In *Proc. IEEE Conf. Comp. Vis. Pattern Recogn.* IEEE, New York.

Quan, L. and Lan, Z. (1999). Linear N-point camera pose determination. *IEEE Trans. Pattern Anal. Mach. Intell.*, 21(8), 774–780.

Ramalingam, S., Lodha, S., Sturm, P. (2004). A generic structure-from-motion algorithm for cross-camera scenarios. In *Proc. 5th Workshop on Omnidirectional Vision, Camera Networks and Non-Classical Cameras*. Center for Machine Perception, Czech Technical University, Prague.

Ramalingam, S., Sturm, P., Lodha, S. (2006). Theory and calibration algorithms for axial cameras. In *Proc. Asian Conf. Comp. Vis.*, volume I. Springer, Heidelberg.

Schröter, H. (1880). *Theorie der Oberflächen zweiter Ordnung und der Raumkurven dritter Ordnung als Erzeugnisse projektivisher Gebilde – Nach Jacob Steiner's Prinzipien auf synthetischem Wege abgeleitet.* B.G. Teubner, Leipzig.

Seitz, S. and Kim, J. (2002). The space of all stereo images. *Int. J. Comput. Vision*, 48(1), 21–38.

Stewénius, H., Nistér, D., Oskarsson, M., Åström, K. (2005). Solutions to minimal generalized relative pose problems. In *Proc. 6th Workshop on Omnidirectional Vision, Camera Networks and Non-Classical Cameras*. IEEE, New York.

Sturm, P., Ramalingam, S., Tardif, J.-P., Gasparini, S., Barreto, J. (2011). Camera models and fundamental concepts used in geometric computer vision. *Foundations and Trends in Computer Graphics and Vision*, 6(1–2), 1–183.

Swaminathan, R., Grossberg, M., Nayar, S. (2004). Designing mirrors for catadioptric systems that minimize image errors. In *Proc. 5th Workshop on Omnidirectional Vision, Camera Networks and Non-Classical Cameras*. Prague.

Terrero, A. (1862). Topofotografía, ó sea aplicaciones de la fotografía al levantamiento de los planos topográficos. *La Asamblea del Ejército y de la Armada*, V/2/3, 31–46.

Torralba, A. and Freeman, W. (2012). Accidental pinhole and pinspeck cameras: Revealing the scene outside the picture. In *Proc. IEEE Conf. Comp. Vis. Pattern Recogn.* IEEE, New York.

Treibitz, T., Schechner, Y., Kunz, C., Singh, H. (2012). Flat refractive geometry. *IEEE Trans. Pattern Anal. Mach. Intell.*, 34(1), 51–65.

Triggs, B., McLauchlan, P., Hartley, R., Fitzgibbon, A. (1999). Bundle ajustment – A modern synthesis. In *Proc. Int. Workshop Vision Algorithms: Theory and Practice*. Springer, Heidelberg.

2

Models and Calibration Methods

Guillaume CARON[1,2]
[1]*MIS Laboratory, University of Picardie Jules Verne, Amiens, France*
[2]*CNRS-AIST Joint Robotics Laboratory (JRL), Tsukuba, Japan*

This chapter reviews the state-of-the-art projection models for omnidirectional cameras and the calibration methods to compute their parameters. These models range from the expression of explicit shape of mirrors and lenses to unified and generic models. The calibration methods estimate the parameters of these models because of partially or totally known scene geometry. Various geometric features are considered for camera calibration such as points, lines, circles used as input of linear estimators or nonlinear optimization schemes.

Several examples of cameras and images they capture are provided together with sketches, illustrating the path of light rays entering cameras as well as the main equations using common notations to ease their understanding and comparison.

2.1. Introduction

This chapter deals with the geometrical modeling of omnidirectional image formation and the methods to compute its parameters for each omnidirectional camera. The models are formalized by relying on the geometry of the optics of each omnidirectional camera, exploiting the explicit shape of the mirrors and lenses, that is, the ad hoc models (section 2.2.2), or by more abstract relations valid for several types of omnidirectional cameras, such as the unified central projection model (section 2.2.3). On the other hand, the generic models (section 2.2.4) make it possible to compensate for the impossibility of characterizing the omnidirectional camera by the other models or the great difficulty in calibrating it because of a too large number of parameters.

Finally, the calibration methods (section 2.3) estimate the parameters of the above-mentioned models from correspondences between the content of the omnidirectional image and the observed scene, which is known either totally by the use of calibration patterns or partially, whether the scene is structured or not.

2.2. Projection models

2.2.1. *Perspective projection: a review*

Describing the geometrical formation of an image (Figure 2.1(c)), observation of a scene by a camera (Figure 2.1(a)) assimilable to a pinhole or a camera lens with an infinite depth of field, the perspective projection model, very well known, involves a projection center $\mathbf{C} \in \mathbb{R}^3$ and the image plane π (Figure 2.1(b)). With the image being a finite rectangle of π, it defines a section of the pyramid of vertex $\mathbf{C}$ and infinite height characterizing the field of view of the camera. The greater the focal length $f \in \mathbb{R}_+^*$ between the optical center and the image plane, the smaller the field of view and vice versa.

We define the camera frame $\mathcal{F}_c$ with origin $\mathbf{C}$ and axes $\mathbf{X}_c \in \mathbb{R}^3$, such that $\|\mathbf{X}_c\| = 1$, and $\mathbf{Y}_c \in \mathbb{R}^3$, such that $\|\mathbf{Y}_c\| = 1$, parallel to the horizontal and, respectively, vertical edges of the image rectangle. By usual convention, the direction of these axes follows that of the organization of the pixels of a digital image acquired by a camera, that is, "to the right" for $\mathbf{X}_c$ and "downward" for $\mathbf{Y}_c, \mathbf{Z}_c \in \mathbb{R}^3$, such that $\|\mathbf{Z}_c\| = 1$, is orthogonal to the previous axes making $\mathcal{F}_c$ a direct reference frame, so $\mathbf{Z}_c$ points to the front of the camera. Thus, the three-dimensional (3D) points $^c\mathbf{X} = [^cX \ \ ^cY \ \ ^cZ]^\mathsf{T} \in \mathbb{R}^3$ of the scene observed by the camera, expressed in the camera reference frame $\mathcal{F}_c$, have their third coordinate positive.

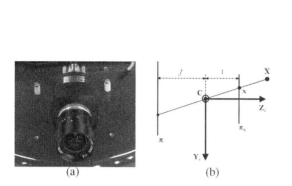

(a)   (b)   (c)

Figure 2.1. *Illustration of the perspective projection. (a) Conventional camera. (b) Diagram of the perspective projection of a point* **X** *of the 3D world in the normalized image plane* π_x. *(c) Example of an image acquired by a camera that can be characterized by the perspective projection model from the triforium of the cathedral of Amiens: despite the exceptional regularity of the building, the width of the central vessel in the image decreases from the bottom of the image toward its center while it is approximately the same along the entire length of the building. For a color version of this figure, see www.iste.co.uk/vasseur/omnidirectional.zip*

The perspective projection model expresses an image point by the intersection of the line of sight $(^cC^cX)$, representing the path followed by a light ray, with the image plane. If, physically, the image plane is beyond the optical center with respect to the 3D point, any plane in space parallel to it can form a virtual image plane, identical to the real image plane, with a similarity transformation. The normalized image plane π_x, that is, distant of one unit from ^cC, on the side of ^cX, is generally chosen in computer vision.

Therefore, the perspective projection model expresses the projection of a 3D point ^cX in π_x as $\mathbf{x} = (x \ y)^\mathsf{T} \in \mathbb{R}^2$ by

$$\begin{cases} x = \dfrac{^cX}{^cZ}, \\[2ex] y = \dfrac{^cY}{^cZ}. \end{cases} \qquad [2.1]$$

x is defined in the normalized image plane whose origin is its own intersection with the $\mathbf{Z}_c$ axis. It is to be noted that

$$^c\mathbf{x} = (x \quad y \quad 1)^\mathsf{T} \in \mathbb{R}^3, \qquad\qquad [2.2]$$

gives the direction of the line of sight, expressed in $\mathcal{F}_c$ and passing through its origin. It is distinguished from $\tilde{\mathbf{x}} = (x \quad y \quad 1)^\mathsf{T} \in \mathbb{P}^2$, the homogeneous representation of x in $\pi_\mathbf{x}$. We then write the perspective projection function $pr()$

$$\tilde{\mathbf{x}} = pr(^c\mathbf{X}), \qquad\qquad [2.3]$$

with x and y expressed as in [2.1].

The origin of the digital image acquired by a camera being at the top left and its sampling being in pixels, the perspective projection model considers an additional affine transformation $\mathbf{K} \in \mathrm{Aff}(2)$ to transform $\pi_\mathbf{x}$ to the digital image plane $\pi_\mathbf{u}$. This transformation involves, generally, four parameters $\gamma_p = \{\alpha_u, \alpha_v, u_0, v_0\}$ of which $\alpha_u \in \mathbb{R}^*$ and $\alpha_v \in \mathbb{R}^*$ are the horizontal and, respectively, vertical scale factors and $(u_0, v_0) \in \mathbb{R}^2$ are the coordinates of the principal point, that is, the intersection of $\pi_\mathbf{x}$ and $\mathbf{Z}_c$, expressed in the digital image. These parameters, called intrinsic, characterize the optics of the camera, according to the perspective projection model and are linked to the physical realization of an image, in particular $\alpha_u = f/k_u$ and $\alpha_v = f/k_v$, with $k_u \in \mathbb{R}_+^*$ and $k_v \in \mathbb{R}_+^*$ the dimensions of a photodiode giving a pixel in the digital image. Thus, the point $\tilde{\mathbf{u}} = (u \quad v \quad 1)^\mathsf{T} \in \mathbb{P}^2$ of the digital image is obtained from $\tilde{\mathbf{x}}$ by

$$\tilde{\mathbf{u}} = \mathbf{K}\tilde{\mathbf{x}} \ \text{ with } \ \mathbf{K} = \begin{pmatrix} \alpha_u & 0 & u_0 \\ 0 & \alpha_v & v_0 \\ 0 & 0 & 1 \end{pmatrix}. \qquad\qquad [2.4]$$

By putting together the two steps [2.3] and [2.4], we obtain the projection function of a 3D point $^c\mathbf{X}$ in the digital image plane

$$\tilde{\mathbf{u}} = pr_{\gamma_p}(^c\mathbf{X}) = \mathbf{K}pr(^c\mathbf{X}). \qquad\qquad [2.5]$$

REMARK 2.1.– *Enrichment of the perspective projection model*: the above projection model takes into account the fact that pixels may not be perfectly square but rectangular. It can be simplified to a minimum with a single scale factor. This model can also be enriched to take into account: the fact that the pixels are parallelograms, radial or tangential distortions caused by the optics used, a misalignment of the optics and the photodiode array of the camera, etc.

The perspective projection model is suitable for cameras with a limited field of view, around 122° without distortion[1]. Indeed, the division by ^cZ in the perspective projection equations [2.1] raises three problems when the field of view approaches or exceeds 180°, for example when using a curved mirror catadioptric lens (Figure 2.2(a)) or a fisheye lens (up to[2] 280° fisheye lens). The first problem, when ^cZ tends to 0, generates the need for a very large image rectangle on $\pi_\mathbf{x}$, either because of a very large photosensitive matrix, or at the expense of the resolution in the center of the image. Then, when $^cZ = 0$, x and y are not defined in [2.1], not to mention the numerical instability of the calculations when ^cZ is very close to 0. Finally, when the field of view is greater than 180°, two 3D points $^c\mathbf{X}$ and $^c\mathbf{X}'$ such that $^cX' = -^cX$, $^cY' = -^cY$ and $^cZ' = -^cZ$ project to the same coordinates in the image plane. This is why the projection function $pr_{\gamma_p}()$ [2.5] must be adapted for omnidirectional cameras by taking into account either explicitly the optical geometry of the lenses (section 2.2.2), or an abstraction replacing the plane surface of the perspective projection model by a sphere (section 2.2.3), or by relaxing the idea of the projection surface by reasoning on a vector field (section 2.2.4).

2.2.2. *Ad hoc models*

Ad hoc projection models explicitly exploit the optical geometry of panoramic and omnidirectional cameras, whether they involve one or more mirrors (section 2.2.2.1), a fisheye lens (section 2.2.2.2) or a combination of cameras (section 2.2.2.3).

----

1 Distortions less than 1% according to www.dxomark.com as of March 1, 2017 for the Sigma 12–24 mm camera lens.
2 Entaniya M12 280.

2.2.2.1. *Catadioptric cameras*

A catadioptric camera combines curved lenses (diopter) and mirrors (catoptric) to acquire a panoramic image (Figure 2.2(c)).

2.2.2.1.1. Central catadioptric cameras

The panoramic catadioptric camera is single view point, or central, when we associate a hyperbolic mirror of revolution to a perspective camera (hypercatadioptric camera, Figure 2.2(a)), a parabolic mirror of revolution to an orthographic camera (paracatadioptric camera), for convex mirrors, or a concave elliptic mirror of revolution to a perspective camera (Baker and Nayar 1999). The single view point is also almost respected by combining a convex paraboloid mirror and a concave spherical mirror with a perspective camera (Figure 2.2(b)). For all these configurations, ensuring the single view point requires a precise relative placement of the perspective or orthographic camera and the mirror(s): the principal axes must be aligned and the optical center of the perspective camera must be coincident with the focus of the conical surface of the mirror (of the spherical mirror in the case of two mirrors mentioned above). Other combinations of mirrors, although more complex to implement, also allow the realization of a single view point panoramic camera (Bruckstein and Richardson 2000; Gonzalez-Barbosa 2004).

(a)   (b)   (c)

Figure 2.2. *Panoramic catadioptric vision. (a) Single catadioptric camera (V-Stone VS-C450MR-TK lens). (b) Dual catadioptric camera (RemoteReality lens) acquiring panoramic; (c) image (Place Saint-Michel, Amiens, 2015). For a color version of this figure, see www.iste.co.uk/vasseur/omnidirectional.zip*

The ad hoc modeling of central catadioptric cameras is based on the equations of the surfaces of each mirror and those of the associated cameras. Thus, when the placement of the mirror with respect to the camera respects the geometrical constraints to ensure the uniqueness of the view point, we express the coordinates of x the point of the normalized image plane from the coordinates of the corresponding 3D point $^c\mathbf{X}$ and the parameters of the mirror equation (Geyer and Daniilidis 2000; Table 2.1) with

$$\rho = \sqrt{{}^cX^2 + {}^cY^2 + {}^cZ^2}. \qquad\qquad [2.6]$$

Mirror	Parameters	Camera	Projection equations	
Parabolic convex	h (semi-latus rectum)	Orthographic	$\begin{cases} x = \dfrac{h^cX}{\rho - {}^cZ} \\[2mm] y = \dfrac{h^cY}{\rho - {}^cZ} \end{cases}$	[2.7]
Hyperbolic convex	h and e (eccentricity)	Perspective	$\begin{cases} x = \dfrac{2eh^cX/\sqrt{4e^2+h^2}}{\frac{2e}{\sqrt{4e^2+h^2}}\rho - {}^cZ} \\[3mm] y = \dfrac{2eh^cY/\sqrt{4e^2+h^2}}{\frac{2e}{\sqrt{4e^2+h^2}}\rho - {}^cZ} \end{cases}$	[2.8]
Ellipsoidal concave	h et e (eccentricity)	Perspective	$\begin{cases} x = \dfrac{2eh^cX}{2e\rho + {}^cZ\sqrt{4e^2+h^2}} \\[3mm] y = \dfrac{2eh^cY}{2e\rho + {}^cZ\sqrt{4e^2+h^2}} \end{cases}$	[2.9]
Parabolic convex and spherical concave	h and r (radius of the spherical mirror)	Perspective	$\begin{cases} x = \dfrac{2h^cX}{r/(\rho - {}^cZ) - h^2(\rho + {}^cZ)/r} \\[3mm] y = \dfrac{2h^cY}{r/(\rho - {}^cZ) - h^2(\rho + {}^cZ)/r} \end{cases}$	[2.10]

Table 2.1. *Ad hoc projection models of central catadioptric cameras (summary of Geyer and Daniilidis (2000) and of the approximation for the two-mirror projection model (Gonzalez-Barbosa 2004))*

2.2.2.1.2. Non-central catadioptric cameras

Even with a camera-mirror pair that can lead to the single view point, when their relative placement is imprecise, the catadioptric camera is non-central (Ieng and Benosman 2006). This is also the case for any combination of mirror and camera shape, other than those discussed in section 2.2.2.1.1.

In this case, the light rays passing through the camera to reach the image plane do not intersect at a single point (Swaminathan et al. 2001) but form a caustic surface (Hamilton 1828).

The explicit formulation of the caustics associated with the shape of a mirror is based on the flux flow model from geometrical optics. The equation of the caustic surface associated with a mirror is obtained by solving the differential equation canceling the Jacobian of a point $\mathbf{F} \in \mathbb{R}^3$ of the caustic with respect to the height ^cZ_m of the mirror surface and the distance between $\mathbf{F}$ and the mirror along the reflected ray. Using the parameters of eccentricity e and semi-latus rectum h of the mirror surface introduced in section 2.2.2.1.1, the mirror surface is then written parametrically (Swaminathan et al. 2001)

$$g(^cZ_m) = \sqrt{\frac{h^2}{4} - h^cZ_m - (e^2 - 1)^cZ_m^2}, \qquad [2.11]$$

leading to the implicit expression of the caustic

$$f(^cZ_m, g) = (e^2 - 1)^cZ_m^2 + g(^cZ_m)^2 + h^cZ_m - \frac{h^2}{4} = 0. \qquad [2.12]$$

If the shape of the mirror is not known, the factors of the parameters of [2.12] are replaced by unknowns to be estimated numerically.

2.2.2.2. Fisheye cameras

A fisheye camera has a hemispherical field of view (180°) or more. As for the panoramic catadioptric camera, the radial distortions encountered in the image acquired by the fisheye camera are not aberrations but are the result of the projection of a sphere on a plane. When the radial distortions $r \in \mathbb{R}$ are symmetrical with respect to the principal point, they are expressed from the polar angle $\theta \in \mathbb{R}$, formed by the line of sight of the point ^cX and the optical axis $\mathbf{Z}_c$, that is, $\theta = \arccos(^cZ/||^cX||)$. Several fisheye projection models exist, mainly the equidistant (Miyamoto 1964) projections

$$r = \theta, \qquad [2.13]$$

and equisolid (Miyamoto 1964)

$$r = 2\sin\frac{\theta}{2}, \qquad [2.14]$$

to compare with the equivalent relationship for a perspective camera (section 2.2.1)

$$r = \tan\theta. \qquad [2.15]$$

The equidistant projection has a regular radial resolution while the equisolid projection has a better resolution at the center than at the edges of the image.

Whatever the model considered among the three previous ones, we express the coordinates of the normalized image point x from r and the second angle $\phi \in \mathbb{R}$ defining the direction of the line of sight, that is, the azimuthal angle $\phi = \arctan(^cY/^cX)$, by

$$\mathrm{x} = r \begin{pmatrix} \cos\phi \\ \sin\phi \end{pmatrix}. \qquad [2.16]$$

However, real fisheye lenses rarely follow perfectly the ideal models of the equations [2.13] and [2.14]. In this case, a more general projection model is to be considered, such as the angular polynomial projection model (Kannala and Brandt 2000). This model expresses the radial distortions r from a polynomial function of the polar angle θ, denoted $r(\theta)$, considering only the odd powers of θ, weighted by coefficients $k_1, k_2, ...$

$$r(\theta) = k_1\theta + k_2\theta^3 + k_3\theta^5 + k_4\theta^7 + k_5\theta^9 + \dots, \qquad [2.17]$$

without loss of generality. Thus, this model has as many additional intrinsic parameters to the perspective projection model as terms considered in the polynomial [2.17], that is, $\gamma_{pa} = \{\alpha_u, \alpha_v, u_0, v_0, k_1, k_2, \dots\}$.

Alternatively, the Cartesian polynomial projection model (Scaramuzza et al. 2006a) reasons by expressing the radial distortions $r(\rho_{u'})$ in the digital image plane, centered at the principal point u_0 in which the coordinates of a point are expressed as $u' = u - u_0$ and $v' = v - v_0$. This model defines $r(\rho_{u'})$ such that the vector $[u', v', r(\rho_{u'})]$ is collinear with the line of sight associated with the 3D point ^cX, that is, by posing $\alpha = \alpha_u = \alpha_v$

$$\frac{\rho}{\alpha}[u', v', r(\rho_{u'})] = {}^cX, \qquad [2.18]$$

where ρ is the norm of ^cX [2.6], and

$$r(\rho_{u'}) = a_0 + a_1\rho_{u'} + a_2\rho_{u'}^2 + \dots + a_N\rho_{u'}^N, \qquad [2.19]$$

such that

$$\rho_{u'} = \sqrt{u'^2 + v'^2}. \tag{2.20}$$

In summary, similar to the angular polynomial projection model, the number of intrinsic parameters of the Cartesian polynomial projection model depends on the number of coefficients considered in the polynomial [2.19]. They are noted as $\gamma_{pc} = \{\alpha, u_0, v_0, a_1, a_2, \ldots\}$.

2.2.2.3. *Multi-camera systems*

The multi-camera systems combine several cameras, identical or not, in a rig with mainly complementary fields of view in order to reach, together, an omnidirectional field of view up to 360°. This type of polydioptric visual sensor goes from the combination of several perspective cameras distributed on the surface of a sphere (Swaminathan and Nayar 1999) (polycamera) to the combination of two back-to-back fisheye lenses, for the most compact (Li 2006). Some products are available on the market for professionals of photography and panoramic and 360 movies (or *virtual reality*), for the first (e.g. LadyBug, Dodeca 2, Insta360Pro), and many consumer products (e.g. Ricoh Theta, Insta360One, Samsung Gear 360, Garmin Virb 360, etc.), for the second.

Generally, a projection model characterizes each camera, so there are as many sets of intrinsic parameters $\gamma_{m,j}$ as there are cameras, m designating the projection model considered and j, the camera index. To these, we add the extrinsic parameters that express the pose $p_{s,j} \in \mathbb{R}^6$ of each camera j, of frame $\mathcal{F}_{c_j}$, in a common coordinates system associated with the polydioptric system, $\mathcal{F}_s$.

Considering $p_{s,j} = \begin{bmatrix} c_j t_s^\mathsf{T} & \mathsf{r}_{s,j}^\mathsf{T} \end{bmatrix}^\mathsf{T}$, such that $c_j t_s \in \mathbb{R}^3$ and $\mathsf{r}_{s,j} \in \mathbb{R}^3$, the axis-angle representation of the rotation matrix $c_j R_s \in SO(3)$ (Ma et al. 2006), we express $c_j M_s \in SE(3)$, the rigid transformation matrix from frame $\mathcal{F}_s$ to frame $\mathcal{F}_{c_j}$ by

$$c_j M_s = \begin{bmatrix} c_j R_s & c_j t_s \\ \mathbf{0}_{1 \times 3} & 1 \end{bmatrix}. \tag{2.21}$$

This modeling of polydioptric system is very similar to that of stereoscopic vision systems.

2.2.3. *Unified central projection and its extensions*

Unlike ad hoc models, unified central projection is based on a mathematical abstraction of the geometry of central omnidirectional camera image formation. It introduces a virtual sphere as an intermediate projection surface between the scene and the image plane (Geyer and Daniilidis 2000; Barreto and Araújo 2001). This sphere is also another image representation space, common to all central cameras.

2.2.3.1. *Unified central projection model*

The unified central projection model (Figure 2.3) can be seen as a generalization of the perspective projection model consisting of adding a preliminary step, which first projects the 3D point $^c\mathbf{X}$ onto a unit sphere (Barreto and Araújo 2001) of center $\mathbf{C}$ in $\mathbf{X}_S \in \mathbb{R}^3$, such that $\|\mathbf{X}_S\| = 1$

$$\mathbf{X}_S = \begin{pmatrix} X_S \\ Y_S \\ Z_S \end{pmatrix} = pr_S(^c\mathbf{X}) \text{ with } \begin{cases} X_S = \frac{^cX}{\rho} \\ Y_S = \frac{^cY}{\rho} \\ Z_S = \frac{^cZ}{\rho} \end{cases}, \ \rho = \|^c\mathbf{X}\|, \quad [2.22]$$

before projecting it on the normalized image plane in x, using a second center of projection $\mathbf{C} \in \mathbb{R}^3$, distant of $\xi \in \mathbb{R}_+$ from the first one, along the axis $\mathbf{Z}_c$ (Barreto and Araújo 2001)

$$\mathbf{x} = pr\left(\mathbf{X}_S + \begin{pmatrix} 0 & 0 & \xi \end{pmatrix}^\mathsf{T}\right) \text{ with } \begin{cases} x = \frac{X_S}{Z_S+\xi} \\ y = \frac{Y_S}{Z_S+\xi} \end{cases}. \quad [2.23]$$

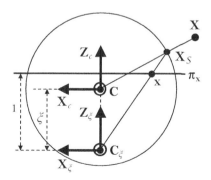

Figure 2.3. *Diagram of the unified central projection model*

$pr_S()$ and $pr()$ combine into a single projection function pr_ξ of the 3D point $^c\mathbf{X}$ into $\tilde{\mathbf{x}}$

$$\tilde{\mathbf{x}} = pr_\xi(^c\mathbf{X}) \text{ with } \begin{cases} x = \frac{^cX}{^cZ+\xi\rho} \\ y = \frac{^cY}{^cZ+\xi\rho} \end{cases}. \qquad [2.24]$$

To finalize the unified central projection model, the transformation from the normalized image plane to the digital image plane is done in the same way as with the perspective [2.4] projection model. ξ joins the set of intrinsic parameters of the unified central projection model $\gamma_u = \{\alpha_u, \alpha_v, u_0, v_0, \xi\}$. Thus, the projection function of a 3D point $^c\mathbf{X}$ in the digital image plane is written as

$$\tilde{\mathbf{u}} = pr_{\gamma_u}(^c\mathbf{X}) = \mathbf{K}pr_\xi(^c\mathbf{X}). \qquad [2.25]$$

The projection $pr_\xi()$ of the sphere to the image plane is invertible, which makes it possible to express a spherical point, and thus the associated line of sight, from an image point

$$\mathbf{X}_S = pr_\xi^{-1}(\mathbf{x}) = \begin{pmatrix} \frac{\xi+\sqrt{1+(1-\xi^2)(x^2+y^2)}}{x^2+y^2+1}x \\ \frac{\xi+\sqrt{1+(1-\xi^2)(x^2+y^2)}}{x^2+y^2+1}y \\ \frac{\xi+\sqrt{1+(1-\xi^2)(x^2+y^2)}}{x^2+y^2+1} - \xi \end{pmatrix}. \qquad [2.26]$$

Equation [2.25] of projection to the digital image plane is equivalently rewritten by introducing the intrinsic parameters $\tau \in [0,1[$, $\alpha'_u \in \mathbb{R}^*$ and $\alpha'_v \in \mathbb{R}^*$ and by posing $\xi = \tau/(1-\tau)$, $\alpha_u = \alpha'_u/(1-\tau)$ and $\alpha_v = \alpha'_v/(1-\tau)$ (Usenko et al. 2018). We then express the equivalent unified central projection function $pr_{\gamma'_u}$

$$\tilde{\mathbf{u}} = pr_{\gamma'_u}(^c\mathbf{X}) = \mathbf{K}'pr_\tau(^c\mathbf{X}) \text{ with } \mathbf{K}' = \begin{pmatrix} \alpha'_u & 0 & u_0 \\ 0 & \alpha'_v & v_0 \\ 0 & 0 & 1 \end{pmatrix}, \qquad [2.27]$$

and $pr_\tau(^c\mathbf{X})$ giving

$$\begin{cases} x = \frac{^cX}{(1-\tau)^cZ+\tau\rho} \\ y = \frac{^cY}{(1-\tau)^cZ+\tau\rho} \end{cases}. \qquad [2.28]$$

This rewriting of the unified central projection model with $\gamma'_u = \{\alpha'_u, \alpha'_v, u_0, v_0, \tau\}$ inverts and has better numerical properties for the calibration (Usenko et al. 2018).

Finally, representing the coordinates of points on the sphere by Cartesian coordinates is redundant because, since the sphere is unitary, X_S, Y_S and Z_S are not independent ($\|X_S\| = 1$). The minimal representation of a point on the sphere is done by the spherical coordinates of azimuthal ϕ and polar θ angles[3] ($\boldsymbol{\theta} = [\theta, \phi]^\mathsf{T} \in \mathbb{R}^2$), which are expressed from $\mathbf{X}_S$ by the function $c2s()$ (Cartesian to spherical)

$$\boldsymbol{\theta} = \begin{pmatrix} \theta \\ \phi \end{pmatrix} = c2s(\mathbf{X}_S) = \begin{pmatrix} \arccos(Z_S) \\ \arctan(Y_S/X_S) \end{pmatrix}. \qquad [2.29]$$

REMARK 2.2.– *Enrichment of the unified central projection model*: similar to the perspective projection model, the unified central projection model can be extended by taking into account additional intrinsic parameters, for example, for radial distortions (Remark 2.1).

REMARK 2.3.– *Generalization of the perspective projection model*: the unified central projection model is valid for any single point of view camera, including perspective cameras. Indeed, it is enough to cancel ξ (or τ) to find back the perspective projection.

2.2.3.2. *Extensions for fisheyes cameras*

The unified central projection model (section 2.2.3.1) is also equivalent to the ad hoc projection models of fisheye lens cameras (Courbon et al. 2012), as discussed in section 2.2.2.2. In practice, this model needs to be completed by a distortion parameter to better approximate the projection of most fisheye lenses (Ying and Hu 2004a). However, this model poorly characterizes fisheye lenses whose field of view is greater than $180°$ (Usenko et al. 2018).

In this case, the unified central projection model can be extended by adding, after the spherical projection [2.22], a second spherical projection, of center $\mathbf{C}_\xi$, before the perspective projection to the image plane of center $\mathbf{C}_\tau = (0, 0, \tau/(1 - \tau) - \xi)^\mathsf{T}$, which thus becomes the third center of

---

3 This chapter considers the standard ISO 80000-2:2019 to note spherical coordinates.

projection. The equation of this double-sphere projection model to the digital image plane is then written from the two versions [2.25] and [2.27] of the unified central projection (Usenko et al. 2018)

$$\tilde{\mathbf{u}} = pr_{\gamma_d}({}^c\mathbf{X}) = \mathbf{K}'pr_\tau(pr_\xi({}^c\mathbf{X})), \qquad [2.30]$$

such that the composition of projection functions $pr_\tau(pr_\xi({}^c\mathbf{X}))$ gives

$$\begin{cases} x = \frac{{}^cX}{(1-\tau)({}^cZ+\xi\rho)+\tau\rho_2} \\ y = \frac{{}^cY}{(1-\tau)({}^cZ+\xi\rho)+\tau\rho_2} \end{cases}, \qquad [2.31]$$

where ρ is defined in equation [2.6], and with

$$\rho_2 = \sqrt{{}^cX^2 + {}^cY^2 + ({}^cZ + \xi\rho)^2}. \qquad [2.32]$$

This last model has therefore six intrinsic parameters $\gamma_d = \{\alpha'_u, \alpha'_v, u_0, v_0, \xi, \tau\}$.

2.2.3.3. *Extension for 360° cameras*

The most compact spherical polydioptric systems have the specificity of being designed so that their lenses have complementary fields of view, thus reducing their number to a minimum. Two fisheye lenses placed back-to-back are enough to cover the full 360° of the spherical field of view because of two flat mirrors judiciously positioned between the lenses, thus reflecting the light to a single photosensitive matrix (Li 2006). For even more compactness, the camera manufacturer Ricoh has replaced, in its Theta[4] brand, the mirrors by two prisms redirecting the light rays passing through the fisheye lenses toward two photosensitive matrices.

Beyond the interest of the very small volume of this spherical polydioptric camera (Figure 2.4(a)), for the general public, as well as in robotics, the proximity of the two fisheye lenses also makes their optical centers very close. We can then make the approximation that they are at the same location, in particular when the elements of the observed scene are far enough from the

---

4 Available at: http://theta360.com.

spherical camera[5] (minimum working distance of polycameras (Swaminathan and Nayar 1999)).

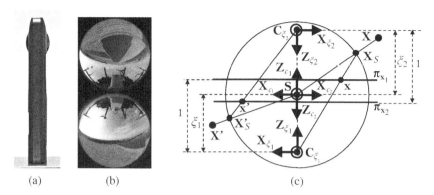

(a)   (b)   (c)

Figure 2.4. *Spherical vision: (a) Compact polydioptric spherical camera (Ricoh Theta S, side view with the two fisheye lenses in its upper part) acquiring (b) double-fisheye images (Le champ à cailloux, Vaux-en-Amiénois, 2019) and whose projection can be represented, under hypotheses, by (c) an extension of the unified central projection model to two image planes. For a color version of this figure, see www.iste.co.uk/vasseur/omnidirectional.zip*

The projection model of a compact polydioptric spherical camera is thus restricted to a single sphere, each hemisphere of which is associated with one of the two fisheye images (Figure 2.4(b)), thus considering two image planes associated with the same sphere (Caron and Morbidi 2018). As the lenses, the photosensitive matrices and their alignment can be slightly different from one fisheye camera to the other, two sets of intrinsic parameters $\gamma_{u,j}$, $j \in \{1, 2\}$ are considered. However, since both fisheye cameras are assumed to share the same origin, we can set $\mathcal{F}_s = \mathcal{F}_{c_1}$ (Figure 2.4(c)) and the extrinsic parameters, that is, the pose of the second fisheye camera, relative to the first, thus, are reduced to the orientation $r_{s,2}$ (or $r_{1,2}$) $^{c_2}R_s = {}^{c_2}R_{c_1} \in SO(3)$. We then re-express the projection of a spherical point $^s\mathbf{X}_\mathcal{S}$ of the hemisphere associated with the camera j in the normalized image plane of the latter (equation [2.23]) by

$$\mathbf{x} = pr_j \left({}^{c_j}R_s {}^s\mathbf{X}_\mathcal{S} + \begin{pmatrix} 0 & 0 & \xi_j \end{pmatrix}^\mathsf{T} \right),$$ [2.33]

5 The Theta brand of Ricoh is mentioned here as an example because it has one of the smallest distances on the market between its fisheyes lenses. It is thus a spherical camera for which the approximation of uniqueness of optical center for the two fisheye lenses is among the most tolerable.

with $^{c_1}R_s = {}^{c_1}R_{c_1} = I_{3\times3}$. In practice, the sign of the third coordinate of $^{c_j}X_s = {}^{c_j}R_s{}^s X_S$ is sufficient to determine which of the two cameras perceives it.

2.2.4. *Generic models*

Generic models, also called discrete models, associate to each pixel a line of sight, also called light ray (Sturm et al. 2011; Ramalingam and Sturm 2017) or raxel (Grossberg and Nayar 2001). These rays intersect at a single point, the optical center, for central cameras. They intersect at the same line, the camera axis, for axial cameras. Finally, the rays are not constrained for the other non-central cameras (Ramalingam and Sturm 2017).

For some axial cameras, the generic model can be quasi-central when the intersection points of the rays and the camera axis form a short segment (Brousseau and Roy 2019). The projection of a 3D point ^cX is then a ray $\psi = [\theta, \phi, \delta_Z]^T \in \mathbb{R}^3$ where θ and ϕ are the polar and azimuthal angles (see [2.29]) of the line of sight formed by ^cX and the point $^c[0, 0, \delta_Z]^T$ (Brousseau and Roy 2019):

$$\psi = \begin{bmatrix} \theta \\ \phi \\ \delta_Z \end{bmatrix} = \begin{bmatrix} c2s\left(\dfrac{^cX - {}^c[0, 0, \delta_Z]^T}{||^cX - {}^c[0, 0, \delta_Z]^T||} \right) \\ \delta_Z \end{bmatrix}. \qquad [2.34]$$

When the camera is central, $\delta_Z = 0$.

δ_Z is possibly unique for each pixel u and the projection model is a table associating to each pixel u a ray ψ.

2.3. Calibration methods

Calibration methods exploit points (most methods), straight lines (Geyer and Daniilidis 2002; Barreto and Araújo 2005; Bermudez-Cameo et al. 2015), circles (Kannala and Brandt 2000) or spheres (Ying and Hu 2004b). These *primitives* observed by the camera on calibration patterns (Figures 2.5a–c) or in the scene (Figure 2.5d) are the data used to estimate the intrinsic parameters of the projection model characterizing the camera. The calibration patterns can be two-dimensional (2D) or 3D, the former requiring it to be observed at several distinct exposures (to be added in the parameters to be estimated) to perform the calibration, when a single observation of a 3D calibration pattern may suffice (Puig et al. 2012). Without requiring a calibration pattern or the presence of a particular structure in the scene, self-calibration relies on the detection and matching of points of interest in the scene in several images acquired at different camera exposures (Kang 2000; Micusik and Pajdla 2006; Nguyen and Lhuillier 2017).

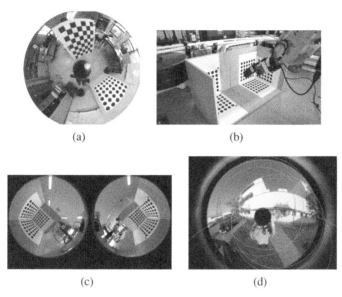

(a)

(b)

(c)

(d)

Figure 2.5. *Examples of measurements in the image for calibration. (a) Checkerboard and disk plane patterns in a paracatadioptric (Caron and Eynard 2011b) image: primitives are points (corners of squares or centers of disks). (b and c) 3D multi-planar pattern observed by a 360 polydioptric camera (Caron and Morbidi 2018). (d) Straight lines extracted from the environment structure (green lines) (Bermudez-Cameo et al. 2015). For a color version of this figure, see www.iste.co.uk/vasseur/omnidirectional.zip*

Most of the available methods and software estimate the intrinsic parameters minimizing the reprojection error (sum of squared differences (SSD)) of reference calibration patterns ${}^{o}X_i$ with respect to their detection in the image u_i

$$\hat{\gamma} = \operatorname*{argmin}_{\gamma} \frac{1}{2} \sum_i ||pr_{\gamma} ({}^{c}M_o {}^{o}X_i) - u_i||^2, \qquad [2.35]$$

$\hat{\gamma}$ being the estimate of the intrinsic parameters according to the considered projection model: $\gamma \in \{\gamma_p, \gamma_u, ...\}$ (section 2.2). In practice, as many ${}^{c}M_o$

as there are calibration patterns must be computed when solving the minimization problem [2.35], even if they are not used afterwards.

As for the solution of the minimization problem [2.35], it is usually iterative by nonlinear optimization (Kannala and Brandt 2000; Usenko et al. 2018; Caron and Eynard 2011b; Mei and Rives 2007; Schönbein et al. 2014a; Caron and Morbidi 2018) by a Newton, Gauss–Newton or Levenberg–Marquardt type method. Indeed, the formation of omnidirectional images is nonlinear. On the other hand, a nonlinear optimization ensures a better management of erroneous or inaccurate measurements (Usenko et al. 2018) than a linear resolution method, being able nevertheless, classically, to initialize the intrinsic parameters to ensure the convergence of the optimization method (Puig et al. 2011a).

Some variants are based on a collinearity criterion, not only when the considered primitives are straight lines (Barreto and Araújo 2005), but also for the lines of sight associated with the pixels of the omnidirectional image, whether they form the projection model (Scaramuzza et al. 2006a; Brousseau and Roy 2019) or not (Puig et al. 2011a).

Table 2.2 gathers the most used and among the most recent omnidirectional camera calibration methods (see Puig et al. (2012) to complete the oldest references). Most of them are associated with free software (Barreto and Araújo 2002; Puig et al. 2011b; Kalibr 2014; Usenko 2018; Caron and Eynard 2011a; Mei 2007; Schönbein et al. 2014b; Scaramuzza 2013) and some of them are even integrated to other libraries like OpenCV[6]. In addition to the attributes already mentioned (use of calibration pattern, type of primitive, criterion considered, solver) to classify the calibration methods, the maximum field of view of the camera calibrated with each method is reported, when known. At most, a monocular camera with a field of view of 280° and a polydioptric camera with a field of view of 360° have been calibrated[7].

---

6 Available at: https://opencv.org.

7 Only polydioptric camera calibration methods leading to an omnidirectional image like monocular are mentioned. For the others, see Liu et al. (2016) and references therein.

Finally, the calibration methods reported in Table 2.2 consider various projection models:

– Ikeda et al. (2003) and Rau et al. (2016) use perspective projection (section 2.2.1) with distortions for multi-camera systems (section 2.2.2.3);

– Li (2006) uses the equidistant fisheye projection (section 2.2.2.2; equation [2.13]);

– models by Kannala and Brandt (2000) and Schönbein et al. (2014a) are based on an angular polynomial projection model (section 2.2.2.2; equation [2.17]), combined with a quadric-based non-central projection model of Agrawal et al. (2011) for Schönbein et al. (2014a);

– the model by Scaramuzza et al. (2006b) is based on the Cartesian polynomial projection model (section 2.2.2.2; equation [2.19]);

– Barreto and Araújo (2005); Puig et al. (2011a); Caron and Eynard (2011b); Mei and Rives (2007) and Caron and Morbidi (2018) use the unified central projection model (section 2.2.3.1);

– the model by Usenko et al. (2018) is based on the double sphere projection model (section 2.2.3.2);

– the model by Brousseau and Roy (2019) is based on a generic projection model (section 2.2.4).

2.4. Conclusion

This chapter has presented the most frequently encountered projection models in omnidirectional vision. Faced with the variety of these models, the choice depends both on the camera used but also on the intended application, particularly according to criteria of accuracy, efficiency and image representation. The calibration method, its ease of implementation and even the availability of software are all practical elements to take into account. Nevertheless, all calibration methods require a distribution of the considered primitives in the whole field of view of the camera, particularly at its periphery, to ensure a correct estimation of the intrinsic parameters of the camera (Puig et al. 2012).

Method	Pattern	Primitive	Criterion	Solver	Max FoV	Software
Barreto and Araújo (2005)	No	Lines	Collinearity	Linear	180[†]	Barreto and Araújo (2002)
Puig et al. (2011a)	Yes	Points	Collinearity and reproj/SSD	Lin and opt/LM	180[†]	Puig et al. (2011b)
Kannala and Brandt (2000)	Yes	Points/circles	Reproj/SCE	Opt/LM	190	Kalibr (2014)
Usenko et al. (2018)	Yes	Points	Reproj/SrCE	Opt/GN	195	Usenko (2018)
Caron and Eynard (2011b)	Yes	Points	Reproj/SCE	Opt/LM	210	Caron and Eynard (2011a)
Mei and Rives (2007)	Yes	Points	Reproj/SCE	Opt/LM	210*	Mei (2007)
Schönbein et al. (2014a)	Yes	Points	Reproj/SCE	Opt	210*	Schönbein et al. (2014b)
Scaramuzza et al. (2006a)	Yes	Points	Collinearity and reproj/SCE	Lin/alt and opt/LM	220	Scaramuzza (2013)
Brousseau and Roy (2019)	Yes	Points (dense)	Collinearity and distance	Opt	280	No
Li (2006); Caron and Morbidi (2018); Ikeda et al. (2003); Rau et al. (2016)	Yes	Points	Reproj/SCE	Opt/LM	360	No

Table 2.2. *Main calibration methods and their characteristics: use of pattern or not, primitive and criterion considered, solver, maximum calibrated field of view (in degrees) and available software. Abbreviations: lin, linear; reproj, reprojection; SSD, sum of squared of differences; rSSD, robust SSD; opt, optimization; LM, Levenberg–Marquardt; GN, Gauss–Newton; alt, alternate. Notes:* [†] *lacks specification;* * *maximum assumed by following that of (Caron and Eynard 2011b) since the methods are similar or the same cameras are calibrated*

Once the camera is calibrated, the representation of the omnidirectional image can be transformed, for example, from the acquired image plane to rectified planes so that the straight lines of the scene become straight and not curved in the image (Barreto and Araújo 2005), or to a spherical image (Makadia and Daniilidis 2006). This issue of omnidirectional image representation depends on the algorithms that exploit it, whether they are video surveillance applications (Tang et al. 2018), 3D reconstructions (Sumikura et al. 2019), virtual reality (Huang et al. 2017), robotics (Caron et al. 2013) or autonomous vehicles (Heng et al. 2019).

2.5. References

Agrawal, A., Taguchi, Y., Ramalingam, S. (2011). Beyond Alhazen's problem: Analytical projection model for non-central catadioptric cameras with quadric mirrors. In *Proc. IEEE Conf. Comput. Vis. Pattern Recognit.*, 2993–3000. Colorado Springs.

Baker, S. and Nayar, S. (1999). A theory of single-viewpoint catadioptric image formation. *Int. J. Comput. Vision*, 35(2), 175–196.

Barreto, J.P. and Araújo, H. (2001). Issues on the geometry of central catadioptric imaging. In *Proc. IEEE Conf. Comp. Vis. Pattern Recogn.*, volume 2. IEEE, Kauai.

Barreto, J.P. and Araújo, H. (2002). CatPack toolbox: Matlab software package for the calibration of Central Catadioptric Cameras using line images [Online]. Available at: https://home.deec.uc.pt/jpbar/CatPack/main.htm.

Barreto, J.P. and Araújo, H. (2005). Geometric properties of central catadioptric line images and their application in calibration. *IEEE Trans. on Pattern Analysis and Machine Intelligence*, 27(8), 1327–1333.

Bermudez-Cameo, J., López-Nicolás, G., Guerrero, J.J. (2015). Automatic line extraction in uncalibrated omnidirectional cameras with revolution symmetry. *Int. J. Comput. Vis.*, 114(1), 16–37.

Brousseau, P.-A. and Roy, S. (2019). Calibration of axial fisheye cameras through generic virtual central models. In *Proc. IEEE Int. Conf. Comput. Vis.* IEEE, Seoul.

Bruckstein, A.M. and Richardson, T.J. (2000). Omniview cameras with curved surface mirrors. In *Proc. IEEE Workshop on Omnidirectional Vision*. IEEE, Hilton Head.

Caron, G. and Eynard, D. (2011a). Hyscas: Hybrid stereoscopic calibration software [Online]. Available at: https://mis.u-picardie.fr/g-caron/software.

Caron, G. and Eynard, D. (2011b). Multiple camera types simultaneous stereo calibration. In *Proc. IEEE Int. Conf. Robot. Automat.* IEEE, Shanghai.

Caron, G. and Morbidi, F. (2018). Spherical visual gyroscope for autonomous robots using the mixture of photometric potentials. In *Proc. IEEE Int. Conf. Robot. Autom.* IEEE, Brisbane.

Caron, G., Marchand, E., Mouaddib, E. (2013). Photometric visual servoing for omnidirectional cameras. *Auton. Robot.*, 35(2–3), 177–193.

Courbon, J., Mezouar, Y., Martinet, P. (2012). Evaluation of the unified model of the sphere for fisheye cameras in robotic applications. *Adv. Rob.*, 26(8–9), 947–967.

Geyer, C. and Daniilidis, K. (2000). A unifying theory for central panoramic systems and practical implications. *Proc. 6th Eur. Conf. Comput. Vis.*, 445–461.

Geyer, C. and Daniilidis, K. (2002). Paracatadioptric camera calibration. *IEEE Trans. Pattern Anal. Mach. Intell.*, 24(5), 687–695.

Gonzalez-Barbosa, J.J. (2004). Vision panoramique pour la robotique mobile : stéréovision et localisation par indexation d'images. PhD Thesis, École doctorale en informatique et télécommunications, Université Toulouse III.

Grossberg, M.D. and Nayar, S.K. (2001). A general imaging model and a method for finding its parameters. In *Proc. IEEE Int. Conf. Comp. Vis.*, volume 2. IEEE, Vancouver.

Hamilton, W. (1828). Theory of systems of rays. *Trans. Royal Irish Academy*, 15, 69–174.

Heng, L., Choi, B., Cui, Z., Geppert, M., Hu, S., Kuan, B., Liu, P., Nguyen, R., Yeo, Y.C., Geiger, A. et al. (2019). Project AutoVision: Localization and 3D scene perception for an autonomous vehicle with a multi-camera system. In *Proc. Int. Conf. Robot. Autom.* IEEE, Montreal.

Huang, J., Chen, Z., Ceylan, D., Jin, H. (2017). 6-DOF VR videos with a single 360-camera. In *Proc. IEEE Virtual Reality*. IEEE, Los Angeles.

Ieng, S. and Benosman, R. (2006). Geometric construction of the caustic curves for catadioptric sensors. In *Imaging Beyond the Pinhole Camera*, Daniilidis, K. and Klette, R. (eds). Springer, Dordrecht.

Ikeda, S., Sato, T., Yokoya, N. (2003). Calibration method for an omnidirectional multicamera system. In *Stereoscopic Displays and Virtual Reality Systems X*, volume 5006. SPIE, Santa Clara.

Furgale, P., Sommer, H., Maye, J., Rehder, J., Schneider, T., Oth, L. (2014). Kalibr: C++ calibration toolbox [Online]. Available at: https://github.com/ethz-asl/kalibr.

Kang, S.B. (2000). Catadioptric self-calibration. In *Proc. IEEE Conf. Comp. Vis. Pattern Recogn.*, volume 1. IEEE, Hilton Head.

Kannala, J. and Brandt, S. (2006). A generic camera model and calibration method for conventional, wide-angle, and fish-eye lenses. *IEEE Trans. Pattern Anal. Mach. Intell.*, 28(8), 1335–1340.

Li, S. (2006). Full-view spherical image camera. In *Proc. IEEE Conf. Pattern Recogn.*, volume 4. IEEE, Hong Kong.

Liu, A., Marschner, S., Snavely, N. (2016). Caliber: Camera localization and calibration using rigidity constraints. *Int. J. Comput. Vision*, 118(1), 1–21.

Ma, Y., Soatto, S., Košecká, J., Sastry, S.S. (2004). *An Invitation to 3D Computer Vision: From Images to Geometric Models.* Springer, New York.

Makadia, A. and Daniilidis, K. (2006). Rotation recovery from spherical images without correspondences. *IEEE Trans. Pattern Anal. Mach. Intell.*, 28(7), 1170–1175.

Mei, C. (2007). Improved omnidirectional calibration toolbox [Online]. Available at: http://www.robots.ox.ac.uk/cmei/Toolbox.html.

Mei, C. and Rives, P. (2007). Single view point omnidirectional camera calibration from planar grids. In *Proc. IEEE Int. Conf. Robot. Automat.* IEEE, Rome.

Micusik, B. and Pajdla, T. (2006). Structure from motion with wide circular field of view cameras. *IEEE Trans. Pattern Anal. Mach. Intell.*, 28(7), 1135–1149.

Miyamoto, K. (1964). Fish eye lens. *J. Opt. Soc. Am.*, 54(8), 1060–1061.

Nguyen, T.-T. and Lhuillier, M. (2017). Self-calibration of omnidirectional multi-cameras including synchronization and rolling shutter. *Comput. Vis. Image Und.*, 162, 166–184.

Puig, L., Bastanlar, Y., Sturm, P., Guerrero, J.J., Barreto, J. (2011a). Calibration of central catadioptric cameras using a DLT-like approach. *Int. J. Comput. Vision*, 93(1), 101–114.

Puig, L., Bastanlar, Y., Sturm, P., Guerrero, J.J., Barreto, J. (2011b). Direct linear transform-like omnidirectional camera calibration [Online]. Available at: http://webdiis.unizar.es/lpuig/DLTOmniCalibration.

Puig, L., Bermúdez, J., Sturm, P., Guerrero, J.J. (2012). Calibration of omnidirectional cameras in practice: A comparison of methods. *Comput. Vis. Image Und.*, 116(1), 120–137.

Ramalingam, S. and Sturm, P. (2017). A unifying model for camera calibration. *IEEE Trans. Pattern Anal. Mach. Intell.*, 39(7), 1309–1319.

Rau, J.Y., Su, B.W., Hsiao, K.W., Jhan, J.P. (2016). Systematic calibration for a backpacked spherical photogrammetry imaging system. *Int. Arch. Photogramm. Remote Sens. Spat. Inf. Sci. – ISPRS Arch.*, XLI-B1, 695–702.

Scaramuzza, D. (2013). Ocamcalib: Omnidirectional camera calibration toolbox for Matlab [Online]. Available at: https://sites.google.com/site/scarabotix/ocamcalib-toolbox.

Scaramuzza, D., Martinelli, A., Siegwart, R. (2006a). A flexible technique for accurate omnidirectional camera calibration and structure from motion. In *Proc. IEEE Int. Conf. Comp. Vis. Syst.* IEEE, New York.

Scaramuzza, D., Martinelli, A., Siegwart, R. (2006b). A toolbox for easily calibrating omnidirectional cameras. In *Proc. IEEE/RSJ Int. Conf. Intel. Robots Syst.* IEEE, Beijing.

Schönbein, M., Strauss, T., Geiger, A. (2014a). Calibrating and centering quasi-central catadioptric cameras. In *Proc. IEEE Int. Conf. Robot. Automat.* IEEE, Hong Kong.

Schönbein, M., Strauss, T., Geiger, A. (2014b). LIBOMNICAL: Omnidirectional camera calibration [Online]. Available at: http://www.cvlibs.net/projects/omnicam/.

Sturm, P., Ramalingam, S., Tardif, J.-P., Gasparini, S., Barreto, J. (2011). Camera models and fundamental concepts used in geometric computer vision. *Foundations and Trends in Computer Graphics and Vision*, 6(1–2), 1–183.

Sumikura, S., Shibuya, M., Sakurada, K. (2019). OpenVSLAM: A versatile visual SLAM framework. In *Proc. ACM Int. Conf. on Multimedia*. ACM, Nice.

Swaminathan, R. and Nayar, S.K. (1999). Polycameras: Camera clusters for wide angle imaging. Technical report CUCS-013-99, Columbia University.

Swaminathan, R., Grossberg, M., Nayar, S. (2001). Caustics of catadioptric cameras. *Proc. IEEE Int. Conf. Comp. Vis.*, 2, 2–9.

Tang, Y., Gao, Z., Lin, F., Li, Y., Wen, F. (2018). Visual adaptive tracking for monocular omnidirectional camera. *J. Vis. Commun. Image R.*, 55, 253–262.

Usenko, V. (2018). The double sphere camera model calibration software (C++) [Online]. Available at: https://gitlab.com/VladyslavUsenko/basalt.

Usenko, V., Demmel, N., Cremers, D. (2018). The double sphere camera model. In *Proc. IEEE Int. Conf. 3D Vision*, 552560.

Ying, X. and Hu, Z. (2004a). Can we consider central catadioptric cameras and fisheye cameras within a unified imaging model. In *Proc. Eur. Conf. Comp. Vis.* Springer, Prague.

Ying, X. and Hu, Z. (2004b). Catadioptric camera calibration using geometric invariants. *IEEE Trans. Pattern Anal. Mach. Intell.*, 26(10), 1260–1271.

3

Reconstruction of Environments

Maxime LHUILLIER
University Clermont Auvergne, CNRS, Institut Pascal,
Clermont-Ferrand, France

The reconstruction of environments from omnidirectional images has two steps. The first one estimates the geometry and is described in the other chapters in this book. It includes SfM, calibration or self-calibration. It estimates all parameters of the camera(s) including six degrees of freedom (DoF) poses, intrinsic parameters and radial distortions. SfM also provides a sparse cloud of 3D points reconstructed from features (e.g. points and lines) detected and matched in the images. The second step provides an approximation of the environment that is more complete than the sparse cloud: a dense cloud of points or a triangulated surface in 3D. This chapter surveys the second step in previous works. In most cases, they assume that the camera is moving in a rigid scene, or similarly, that several cameras take images of a non-rigid scene at a same time.

We start with prerequisites on reconstruction using perspective cameras (section 3.1) and a discussion on omnidirectional cameras for reconstruction (section 3.2). Then previous works are classified in several groups: dense stereo adapted to omnidirectional cameras (section 3.3), reconstruction from only one central image (section 3.4), reconstruction of a non-rigid scene by using a stationary non-central camera (section 3.5) and reconstruction by a moving camera (section 3.6). Finally, we conclude the chapter in section 3.7.

Omnidirectional Vision,
coordinated by Pascal V ASSEUR and Fabio M ORBIDI. © ISTE Ltd 2023.

3.1. Prerequisites

Section 3.1 mostly summarizes standard methods for reconstruction of a rigid scene from input images taken by a perspective camera. This is helpful to understand the case of an environment reconstructed using an omnidirectional camera. We start with reminders on image rectification and matching constraints (section 3.1.1), links between disparity and depth (section 3.1.2). Then dense stereo methods are summarized: semi-global matching (section 3.1.3), plane sweeping (section 3.1.4), minimization of global energy (section 3.1.5) and propagation methods (section 3.1.6). There are also surface reconstruction methods from point clouds (section 3.1.7). All these methods are used or adapted to omnidirectional images in the works described in the remainder of the chapter. Lastly, section 3.1.8 briefly mentions that 3D can be provided by other sensors added to an omnidirectional camera.

3.1.1. *Image rectification and matching constraints*

If two points in two distinct images are projections of a same 3D point in the scene, they meet the epipolar constraint (Hartley and Zisserman 2000): there is a pair of corresponding epipolar lines that include the two points. Since this constraint is known by the geometry estimation step, the matching problem of a pixel in one image is reduced to a 1D search for another pixel in the other image. This search is simplified and accelerated by *rectification*: there are 2D homographies that map the original images to auxiliary images, named rectified images, such that every pair of corresponding epipolar lines become a same horizontal line. Such a rectification cannot be applied in a neighborhood of the epipoles, if they are in the original images (otherwise, the rectified images would be infinite).

After rectification, we can search for the match of a point $(x \ y)$ in the first image as a point $(x + d \ y)$ in the second image by finding the disparity d that minimizes a cost: a photo-consistency measure like the absolute difference of gray levels of the two points (x, y and d are integers). The robustness of this cost increases by *spatial aggregation*, for example, if we replace it by the sum of absolute differences in local windows centered at $(x \ y)$ and $(x + d \ y)$. Other costs are possible, like the sum of squared differences, one minus zero-mean centered correlation, using luminance or RGB channels, etc.

The strategy of computing the disparity d of each pixel as a simple search for the minimizer of an aggregated photo-consistency measure is called winner takes all (WTA). However, there are two cases where the minimizer is not well defined and WTA is unreliable. In the former, the pixel has low texture in its neighborhood. In the latter, which is called the aperture problem, the image gradient is orthogonal to the epipolar line in the search area of the disparity computation. These cases occur even if the search space is reduced by enforcing lower and upper bounds on d. Thus, additional constraints are needed to reduce the risk of bad matches, for example

– disparity gradient limit: disparities $d(x)$ and $d(x + 1)$ of adjacent pixels at x and $x + 1$ meet $|d(x) - d(x + 1)| \leq c$, where c is a threshold;

– ordering: the matching conserves the ordering, that is, $x + d(x) < x' + d(x')$ if $x < x'$;

– uniqueness: the function $x \mapsto x + d(x)$ is injective.

This list of constraints is not exhaustive.

3.1.2. *From disparity to depth*

Since the goal is reconstruction, section 3.1.2 reminds us about the relations between disparity d and depth Z. Assume for a moment that we have a pair of rectified images (section 3.1.1) with disparities computed by dense stereo, for example, a method explained in the following paragraphs. The projections of a 3D point $(X\ Y\ Z)$ in the two rectified images are

$$(x\ y) = (fX/Z\ fY/Z) \quad \text{and} \quad (x + d\ y) = (f(X + T)/Z\ fY/Z)$$

where f is the focal length expressed in pixels and T is the baseline. Thus, the depth is the inverse of the disparity up to a scale factor

$$Z = \frac{fT}{d}. \qquad\qquad [3.1]$$

Furthermore, we examine how a small error in the disparity propagates to an error of the depth. Assume that d follows a random variable with mean d_0 and standard deviation σ_d. By using a first-order approximation of Z as a function of d, Z follows a random variable with mean $Z_0 = fT/d_0$ and standard deviation σ_Z that meets $\sigma_Z^2 = (\frac{\partial Z}{\partial d}(d_0))^2 \sigma_d^2$. We obtain

$$\sigma_Z = \frac{Z_0^2}{f|T|}\sigma_d. \qquad\qquad [3.2]$$

On the one hand, we see that experimental choices provide depths with small errors: large image resolution (i.e. large f), large baseline $|T|$ and scenes without far background (i.e. small Z_0). On the other hand, camera price increases with large resolution, disparity computation is more difficult with large baseline, and reconstruction can be restricted to foreground. Thus, there must be a compromise.

3.1.3. *Dynamic programming and semi-global matching methods (SGM)*

A lot of methods are possible to enforce matching constraints. Dynamic programming minimizes a sum of pixel costs for each epipolar line by enforcing several constraints: uniqueness, ordering and bounds. It also uses the epipolar constraint by using two rectified images. It first computes a cost $C(x_1, x_2)$ in the disparity space image, that is, for all x_1 in the first image and x_2 in the second one, with bounds on $x_2 - x_1$ where $x_i \in \mathbb{N}$. Then, it searches the connected path in the disparity space image from $(0, 0)$ to (x_{max}, x_{max}) that minimizes the sum of C along the path corrected by terms to deal with occlusions. The path only has moves $(+1, +1)$, $(+1, 0)$ and $(0, +1)$ to enforce the constraints and make a recursive computation of the sum possible. The move $(+1, +1)$ corresponds to a match between pixel pairs and the two other moves correspond to occlusions.

However, dynamic programming does not enforce disparity constraint between adjacent lines and makes errors for this reason. The popular SGM (Hirschmuller 2005) removes this drawback and also considers minimal paths, but in a different way. It first computes a *cost volume* $C(\mathbf{p}, d)$ where $\mathbf{p}$ is the pixel coordinate in a reference image and d is a disparity (d is bounded). Then for each discrete image direction $\mathbf{r}$, for example, $\mathbf{r} \in \{-1, 0, +1\}^2 \setminus \{(0, 0)\}$, it searches the path in the volume with direction $\mathbf{r}$ from image boundaries to every $(\mathbf{p}, d)$ that minimizes the sum of C along the path corrected by terms to deal with occlusion and slanted surfaces. All paths with a given $\mathbf{r}$ are computed recursively: the cost $L_{\mathbf{r}}(\mathbf{p}, d)$ of the minimal path that ends at $(\mathbf{p}, d)$ is a function of $L_{\mathbf{r}}(\mathbf{p} - \mathbf{r})$. Lastly, it chooses the d for each $\mathbf{p}$ that minimizes $\sum_{\mathbf{r}} L_{\mathbf{r}}(\mathbf{p}, d)$. This sum of costs is a spatial aggregation and improves robustness. It is better than that of the dynamic programming, which is restricted to the horizontal direction. The time complexity is the product of the three dimensions of the volume.

SGM can be generalized to more than two images even if they are not rectified. The range of disparity is replaced by a finite set H of hypothesis depths. A cost in the volume is defined for every pair $(\mathbf{p}, Z)$ of pixel $\mathbf{p}$ in a reference image and $Z \in H$ by a multi-view photo-consistency of the reprojections in all images of the 3D point corresponding to $(\mathbf{p}, Z)$. However, the time computation of the cost volume can be long since the time complexity is the product of the volume dimensions and the (squared) number of images.

3.1.4. *Plane sweeping methods*

The plane sweeping methods are other popular (WTA) stereo methods. They estimate a depth map in a reference image by sweeping the scene using parallel planes. They use the following property for fast computations of the photo-consistency: each plane π induces 2D homographies H_i^π between the input images such that, for every 3D point in π, its projection in the reference image is mapped by H_i^π to its projection in the ith image.

During the computation, each pixel $\mathbf{p}$ of the reference image has a depth $Z(\mathbf{p})$ and a photo-consistency $C(\mathbf{p})$ initialized by $C(\mathbf{p}) = +\infty$. For each plane π, photo-consistencies of 3D points in π are computed by mapping all images to the reference image using $\forall i, H_i^\pi$, followed by spatial aggregations. Let $Z_\pi(\mathbf{p})$ be the depth of the 3D point in π that projects to $\mathbf{p}$. Let $C_\pi(\mathbf{p})$ be the photo-consistency of this 3D point. If $C_\pi(\mathbf{p})$ is better (smaller) than $C(\mathbf{p})$, then the method resets $C(\mathbf{p}) = C_\pi(\mathbf{p})$ and $Z(\mathbf{p}) = Z_\pi(\mathbf{p})$. Once this is done for each pixel $\mathbf{p}$, the next plane π is considered to improve the depth map Z.

These computations are ideal for segments of the scene surface that are parallel to the planes. Basic plane sweeping only uses one plane direction: the planes are fronto-parallel. Advanced plane sweeping uses several directions that are the expected in the scene (Gallup et al. 2007), for example, horizontal planes for ground surface and vertical planes for walls.

3.1.5. *Minimization of global energy (or cost function)*

A standard approach for computing a depth map $Z(\mathbf{p})$ for the pixels $\mathbf{p}$ in a reference image is global energy minimization (see section 3.1.3 in Furukawa and Hernandez (2015)). It enforces depth smoothness in a more general way than SGM and the plane sweep methods. It also takes a finite set $H_\mathbf{p}$ of hypothesis depths for each pixel $\mathbf{p}$ as input, but it does not simply assume that the true depth has the best photo-consistency. Here, the depth-per-pixel depends on those in the immediate neighborhood. The solution Z minimizes a cost function

$$E(Z) = \sum_{\mathbf{p}} C(\mathbf{p}, Z(\mathbf{p})) + \lambda \sum_{(\mathbf{p},\mathbf{q}) \in \mathcal{N}} R(Z(\mathbf{p}), Z(\mathbf{q})) \qquad [3.3]$$

where $Z(\mathbf{p}) \in H_{\mathbf{p}}$, $C(\mathbf{p}, Z)$ is a photo-consistency measure of the 3D point with depth Z projected in $\mathbf{p}$, R is a spatial regularization term, $\sum_{\mathbf{p}}$ is a sum for all pixels in the reference image, $\sum_{(\mathbf{p},\mathbf{q}) \in \mathcal{N}}$ is a sum for all pairs of adjacent pixels in the reference image using a neighborhood system (4-neighborhood or 8-neighborhood), and $\lambda > 0$ is a weighting parameter. The spatial regularization tends to enforce similar depths for adjacent pixels in the solution Z (the greater λ is, the stronger the regularization). In practice, better results are obtained by bounding C and R, for example, use

$$R(Z_1, Z_2) = \min(\tau, |Z_1 - Z_2|) \qquad [3.4]$$

with a threshold $\tau > 0$. The terms $\sum_{\mathbf{p}} C$ and $\sum_{(\mathbf{p},\mathbf{q}) \in \mathcal{N}} R$ are also called the data term and smoothing term, respectively.

The minimization is difficult (NP-hard) in a general context, but, fortunately, there are efficient methods to implement it for dense stereo (if R meets the triangular inequality and $R(Z, Z) = 0$) that are based on graph-cut. We can also use a second-order prior smoothness instead of a first-order prior one, for example, use $|Z_1 + Z_3 - 2Z_2|$ instead of $|Z_1 - Z_2|$ for three adjacent pixels instead of two. This reduces the risk of fronto-parallel surface (i.e. with constant depth) and provides a more piece-wise planar surface, but the minimization method is more complicated.

3.1.6. *Propagation methods*

The summarized methods above have drawbacks: they assume a finite set of hypothesis depths or disparities for each pixel and require a post-processing to obtain a global 3D model by merging the depth maps of several images.

Patch propagation (see section 3.2 in Furukawa and Hernandez (2015)) is a standard approach to avoid these drawbacks that directly reconstructs 3D points by using all images. A patch p is a small piece of plane in 3D that locally approximates the scene surface. It is reconstructed by optimizing its photo-consistency, which is a function of the patch center $\mathbf{c}(p)$ and patch normal $\mathbf{n}(p)$, in selected images $V(p)$. First, a set S of seed patches is initialized by matching features in the images and reconstructing them (this can be provided by the geometry estimation step). A new patch will be initialized from the seed patch in its neighborhood. This neighborhood is defined by the adjacency of cells in images that project the patches. Each image cell also stores a list of patches that are projected in the cell. Then, S evolves as follows: pull a patch p from S and try to add new patches p' in its neighborhood. If there is room for p' in the neighborhood of p and if the reconstruction of p' is successful, then p' is added to S and to the image cells where it is projected. In a few words, p' is initialized using $\mathbf{n}(p') = \mathbf{n}(p)$, $V(p') = V(p)$ and $\mathbf{c}(p')$ by ray-plane intersection, then $\mathbf{n}(p')$ and $\mathbf{c}(p')$ are refined by optimizing the photo-consistency of p'. Once S is empty, a filtering removes inconsistent patches (in cells) by examining the visibility and geometry of the patches. The growing and filtering alternate many times, and the patches progressively cover the surface scene.

Match propagation (Lhuillier and Quan 2002) is an ancestor of the patch propagation for two images, which is still used for reconstruction of environments. It also needs neither hypothesis depths nor cost volume definition. The matches are pairs $(\mathbf{p}, \mathbf{q})$ of pixels that have good photo-consistencies and should meet disparity constraints: uniqueness and small 2D disparity gradient (the epipolar constraint is optional). The 2D disparity of $(\mathbf{p}, \mathbf{q})$ is $\mathbf{q} - \mathbf{p}$. The 2D disparity gradient between $(\mathbf{p}, \mathbf{q})$ and $(\mathbf{p}', \mathbf{q}')$ is small if

$$(\mathbf{q}' - \mathbf{p}') - (\mathbf{q} - \mathbf{p}) \in \{-1, 0, +1\}^2. \qquad [3.5]$$

First, a set S of seed matches is initialized from the sparse matching of interest points done by the geometry estimation step. The list L of accepted matches is initialized to empty. Then, S evolves as follows: pull a match $(\mathbf{p}, \mathbf{q})$ from S and try to add new matches $(\mathbf{p}', \mathbf{q}')$ in its neighborhood, that is, if the absolute coordinates of $\mathbf{p}' - \mathbf{p}$ and $\mathbf{q}' - \mathbf{q}$ are below a threshold. If the uniqueness constraint is met (i.e. $\mathbf{p}'$ and $\mathbf{q}'$ do not appear in L) and if the 2D disparity gradient is small, then $(\mathbf{p}', \mathbf{q}')$ is added to S and L. The match $(\mathbf{p}, \mathbf{q})$ has the best photo-consistency in S, which is defined by zero-mean normalized cross-correlation in local windows centered on $\mathbf{p}$ and $\mathbf{q}$. This choice provides a best-first match propagation with advantages including robustness with respect to false seed matches and occlusion handling. The time complexity is only $\mathcal{O}(n \log n)$, where n is the number of pixels in the images.

3.1.7. *Surface reconstruction methods*

The previously summarized methods estimate disparity and depth maps in reference images, points and patches in 3D. They do not provide a triangulated surface, although this is the standard representation for a lot of applications. Thus, we briefly explain how to estimate a triangulated surface using an input cloud of 3D points reconstructed from images.

We focus on methods that discretize the space by using a 3D Delaunay triangulation whose vertices are points. They label each Delaunay tetrahedron with "inside" or "outside", such that the target surface separates the inside and outside tetrahedra. An alternative of tetrahedra would be voxels in a regular grid, but this is not efficient for large-scale environments and uneven distributions of input points. Here, we need definitions. A *reconstruction ray* is a line segment $\mathbf{CP}$ where $\mathbf{P}$ is a point and $\mathbf{C}$ is a location (or center) of a camera that reconstructs $\mathbf{P}$. The full line that includes $\mathbf{C}$ and $\mathbf{P}$ is written $(\mathbf{CP})$. Let O be the set of the outside tetrahedra. The surface ∂O is the *boundary* of O, that is the set of Delaunay triangles that are faces of exactly one tetrahedron in O. The methods summarized in section 3.1.7 enforce visibility constraints: each triangle in ∂O should not be crossed by a reconstruction ray (unless $\mathbf{P}$ is a vertex of the triangle). They also regularize ∂O and deal with noisy input point clouds.

A graph-cut (Vu et al. 2011) method estimates the tetrahedron labels that minimize a cost function $E_{vis} + \lambda E_{qual}$ composed of a visibility term E_{vis} and a surface quality term E_{qual} with a weighting parameter $\lambda > 0$. The term E_{vis} not only avoids a triangle in ∂O that is crossed by a $\mathbf{CP}$, it also enforces a tetrahedron to be in O if it includes a camera location $\mathbf{C}$, and it tries to enforce a tetrahedron to not be in O if it is crossed by a $(\mathbf{CP}) \setminus \mathbf{CP}$, where $\mathbf{P}$ is one of the four vertices of this tetrahedron (Figure 3.1). The term E_{qual} tries to avoid a triangle in ∂O that is a face of a "too flat" tetrahedron. The choice of λ is a trade-off: thin structures (e.g. posts) of ∂O are removed if λ increases, ∂O is less robust to bad points if λ decreases.

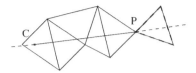

Figure 3.1. *Visibility constraints for surface reconstruction in a 3D Delaunay triangulation using camera center* **C** *and point* **P**. *Here, tetrahedra and triangles are drawn by triangles and edges, respectively. The tetrahedron that includes* **C** *must be in O. The tetrahedron with hatched edges should not be. The four triangles crossed by* **CP** *should not be in* ∂O

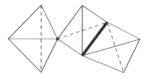

Figure 3.2. *A set of three tetrahedra whose boundary is not manifold. There is a non-manifold vertex at the intersection of the tetrahedra of the left and middle, and a non-manifold edge at the intersection of the tetrahedra of the middle and right (the non-manifold points have coarser pencil stroke). All other boundary points are manifold*

In contrast to the graph-cut methods, a manifold method (Lhuillier and Yu 2013) enforces the manifold constraint on ∂O. This means that every point of the surface (i.e. in $\cup \partial O$) has a surface neighborhood that is homeomorphic to a disk. This constraint is useful not only for surface regularization during computation, but also for post-processing and applications such as surface smoothing, texturing and rendering. Figure 3.2 shows tetrahedra configurations that are avoided by the manifold constraint. Let $r(\Delta)$ be the number of reconstruction ray(s) that cross a Delaunay tetrahedron Δ. We say that Δ is *free-space* if $r(\Delta) > 0$, otherwise Δ is *matter*. The goal is to find the set O of free-space tetrahedra that maximizes a visibility score function

$$r(O) = \sum_{\Delta \in O} r(\Delta) \qquad\qquad [3.6]$$

subject to the constraint that ∂O is manifold. A greedy method approximates the solution of this problem: add progressively free-space tetrahedra in O as long as ∂O remains manifold. First try to add Δ with the highest $r(\Delta)$ such that O grows from the most confident free-space to the less confident free-space. Then try to add several tetrahedra at once to allow topological changes of ∂O and escape from local extrema. Efficient manifold tests are locally applied at each tried update of O.

3.1.8. *Estimation of the 3D using other sensors*

Since this chapter is limited to reconstruction using cameras, it does not detail previous methods that use other sensors to obtain the 3D, even if the acquisition hardware includes an omnidirectional camera. Here are two examples of such hardware. The first one is designed for outdoor environments (Anguelov et al. 2010): Google Street View uses both omnidirectional cameras and laser scanners mounted on the roof of a car. The scanner is used to estimate the dominant planes of the environments. Global positioning systems (GPS) and inertial measurement units (IMU) are also used for the estimation of the camera parameters. Another one is designed for indoor environments (Gokhool et al. 2015): a RGB-D acquisition system is formed by height Asus Xtion Pro Live sensors (Kinect-like sensors). Each of them has a structured light depth sensor (it projects a pattern of light that is only seen by an infrared camera) and a standard camera. They are rigidly fixed around a symmetry axis such that their standard cameras form a $360°$ camera.

3.2. Pros and cons for using omnidirectional cameras

Using an omnidirectional camera to reconstruct an environment seems like a good idea. Its field-of-view (FoV) is larger than that of a standard camera: only one image is enough to capture the (almost) complete part of the environment that is visible from a single view point. Thus, the number of images needed for the environment reconstruction is lower than that of a standard camera. This not only simplifies and accelerates the image acquisition (no need to rotate a camera to see all around), but also can accelerate the computation (e.g. the number of camera poses refined by bundle adjustment is smaller) and simplify sub-problems (e.g. the detection of loop closures). Furthermore, this can also improve the accuracy of a reconstructed point since this point is visible and tracked in a higher number of images. However, these qualitative comparisons only hold if the resolutions (number of image pixels per steradian) of the compared omnidirectional and standard cameras are similar. They are, if the omnidirectional camera is a multi-camera, which is formed by a rigid set of several standard cameras that point in various directions. They are not, if the omnidirectional camera is a catadioptric camera, which is formed by a mirror in front of a standard camera. Here, "standard" means monocular and can be fisheye for multi-cameras.

3.2.1. *Multi-cameras*

The multi-camera case (Figure 3.3 (b), (c), (d)) is the most favorable one for reconstruction accuracy, but has drawbacks: standard cameras must be synchronized. Each of them has its own intrinsic parameters including distortion ones, and there are additional parameters of relative poses between them. Furthermore, a lot of reconstruction methods take composite images as input, for example, spherical images, obtained by stitching the images taken by the standard cameras. These methods implicitly assume that the multi-camera is central and its parameters are accurately known for stitching. However, these assumptions can be bad, for example, the larger the number of standard cameras, the more doubtful the central assumption. The consequences are not only blur/ghosting artifacts in the spherical images, but also inaccurate spherical calibration (a perfect spherical image is equirectangular: the spherical coordinates of ray direction corresponding to pixel are linear to pixel coordinates). Bad assumptions obviously degrade the reconstruction accuracy if the spherical images are used instead of the original ones.

3.2.2. *Catadioptric cameras*

The catadioptric case is less favorable for reconstruction accuracy due to its lower resolution. Furthermore, this resolution is not spatially-uniform and the vertical FoV, assuming that the optical axis is vertical, is limited. These drawbacks can be partially compensated by using still images (instead of video images which have lower resolutions) and by using an equiangular catadioptric camera. Equiangular means that the angle between the back-projected ray and optical axis is linear to the distance between the corresponding pixel and the principal point. Equiangular catadioptric cameras also have favorable vertical FoV compared to other catadioptric ones: about 50 degrees above and below the horizontal plane (Figure 3.3(a)). Such considerations are important in practice, for example, to reconstruct both ground surface and facades in an urban environment. However, the equiangular catadioptric cameras are not central cameras, which can complicate the 3D computations. For example, image rectification for depth computation requires the central assumption. An advantage of the usual (central) catadioptric cameras is the small number of intrinsic parameters to be estimated in comparison to the multi-cameras.

(a) Catadioptric camera in Yu and Lhuillier (2012)

(b) Professional rig in Litvinov and Lhuillier (2013)

(c) DIY rig in Lhuillier (2018b)

(d) DIY rig in Nguyen and Lhuillier (2017)

Figure 3.3. *A few omnidirectional systems (among many others) used for environment reconstruction. The catadioptric camera is formed by the 0-360 mirror and the Canon Legria HSF10. It is equiangular with a good vertical FoV for both ground and buildings. The professional rig is a PointGrey's Ladybug multi-camera mounted on a vehicle. It is composed of six global shutter fisheye cameras (the others in the figure are rolling shutter). The do-it-yourself (DIY) rigs are formed by four GoPro Hero 3 cameras mounted on a helmet. For a color version of this figure, see www.iste.co.uk/vasseur/omnidirectional.zip*

3.2.3. *Toward a wide use of the 360° cameras*

Recently, 360° cameras become popular for several reasons: decreasing prices and dimensions, increasing resolutions and frequencies, trendy applications such as 360° videos and content generation for VR. Their growth has been almost explosive between 2015–2020, since more than 30 360° cameras appeared during in this period (Wikipedia n.d.). This (roughly) began when YouTube started to support 360° videos. A lot of them are multi-cameras composed of two fisheyes that see in opposite directions (Figure 3.4), with a price less than 800 euros. For these reasons, and because of their resolution advantage over catadioptric cameras, the bi-cameras became dominant in the publications about reconstruction using omnidirectional cameras after 2015. Before 2015, most publications used catadioptric cameras or professional (costly) multi-cameras.

Figure 3.4. *Bi-cameras among many others. From left to right: Ricoh Theta S, Samsung Gear 360, Garmin Virb 360, image pair taken by the Virb. The bi-cameras are dominant in the publications about reconstruction using omnidirectional camera between 2015–2020. For a color version of this figure, see www.iste.co.uk/ vasseur/omnidirectional.zip*

3.3. Adapt dense stereo to omnidirectional cameras

This section surveys methods that adapt dense stereo methods in section 3.1 to images taken by an omnidirectional camera, assuming that the calibration (mapping between ray directions and pixels of the original images) is known. The adaptation is partly in the rectification step (section 3.1.1), with three possibilities in the omnidirectional context: planar, cylindrical and spherical rectifications. Here, the name is that of a coordinate system for ray directions. The pixel color of a rectified image is generically defined as follows: take the 2D coordinates of the pixel, multiply them by scale factors and add offsets to obtain the two coordinates of a ray direction, obtain the color that projects the ray direction in the original image. The standard rectifications transform two original images into two rectified images, such that each pair of corresponding epipolar curves in the former is mapped to a same row (or equivalently, a same column) in the latter. There are also weak versions of rectifications that drop the constraint on the straight epipolar lines and only reduce the distortions between the original images (to make pixel matching easier).

Sections 3.3.1, 3.3.2 and 3.3.3 describe previous works that used spherical, cylindrical and planar rectifications, respectively. Plane sweeping methods are also adapted into sphere sweeping methods and summarized in section 3.3.4. Lastly, there are other methods that use neither sweeping nor standard rectification in section 3.3.5.

3.3.1. *Spherical rectifications*

The method in Li (2006) rectifies images taken by two arbitrary cameras with centers C and C', such that the epipolar curves become horizontal lines (Figure 3.5(a)). It uses spherical coordinates of a 3D point P in two well-chosen spherical coordinate systems: with origins C and C', with the same zenith direction $z = C' - C$ and same azimuth direction a (orthogonal to z). Remember that the latitude $p \in [0, \pi]$ is the angle between z and $P - C$, and the longitude $t \in [0, 2\pi]$ is the angle between a and $P - C$ projected on a plane orthogonal to z. Thus, P has a same longitude t and distinct latitudes p and p' in the two coordinate systems. A rectified image is $L \times 2L$, such that pixel (x, y) has the color of the point in an original image whose back-projected P has coordinates $p = \pi x / L$ and $t = \pi y / L$. Corresponding epipolar curves in the original images (not shown here) are mapped to same horizontal line. Large image distortions occur near boundaries $x = 0$ and $x = L$ where the epipoles are mapped.

The spherical rectification in Li (2006) is applied on a catadioptric camera with parabolic mirror in Arican and Frossard (2007) before a graph-cut method with first-order prior smoothness (section 3.1.5). However, the depths corresponding to the disparities near the epipoles are unreliable, at least because the reconstruction rays corresponding to matched pixels near the epipoles are almost parallel. To improve these depths, a third catadioptric image is taken at another camera location C'' that is not collinear to the first two, C and C', then, a second disparity map is computed between C' and C'' using the same method, lastly, the inaccurate depths of the pair (C, C') near epipoles are replaced by more accurate depths of the pair (C', C'') in the common reference image (of C').

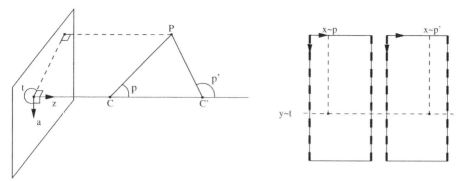

(a) Spherical rectification using azimuth (t) and zenith (p and p') angles

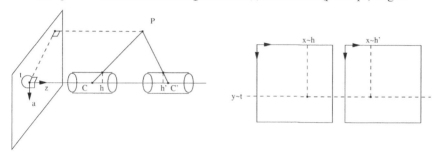

(b) Cylindrical rectification using azimuth angle (t) and heights (h and h')

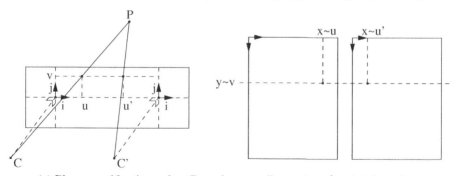

(c) Planar rectification using Cartesian coordinates (u, u' and v) in a plane

Figure 3.5. *Rectifications such that the pairs of corresponding epipolar curves become the same rows. Left: Parameters of rays* $\mathbf{CP}$ *and* $\mathbf{C'P}$. *Right: the x and y coordinates in the rectified images are linear to the ray parameters. Top: the epipoles are projected on the hatched borders. Middle: the two cylinders have the same radius and h is relative to* $\mathbf{C}$ *(h' is relative to* $\mathbf{C'}$). *Bottom: the planar rectification is also given for comparison*

Two rectified spherical images (Li 2006) with a vertical baseline are taken by using a commercial line-scan camera in Kim and Hilton (2013). Then, a global energy function (section 3.1.5) is minimized to estimate a depth map by using a variational approach that avoids disparity discretization. Both the input images with large resolution 12574×5658 and the estimation of floating-point disparities reduce the quantization effects (Bartczak et al. 2007) of the reconstructed surface. Furthermore, a lot of scene details can be reconstructed from only two images thanks to the high resolution and a non-negligible baseline (60 cm). The energy functional has a data term (squared difference of gray levels) and a smoothing term. The latter is anisotropic (it prefers depth discontinuities along image edges) for sharp depth discontinuities and makes it possible to solve the minimization problem by a gradient descent. The former is weighted to deal with occluded areas. Lastly, a hierarchical method is used to escape from local extrema and accelerate the calculations for large images. This algorithm is also used in Thatte et al. (2016) to estimate depth maps in a case where the two images are taken by two spherical cameras (Ricoh Theta S) with a vertical baseline and the same vertical (symmetry) axis.

Urban environments are reconstructed using two catadioptric cameras mounted on a car roof (Schönbein and Geiger 2014). Four images are considered for reconstruction: those of the left (L) and right (R) cameras taken at two consecutive time steps t and $t + 1$. There are three steps once the four view geometry is estimated. First, a depth map is computed using SGM (section 3.1.3) for each pair of rectified spherical images (Li 2006) among pairs L&R at t, L&R at $t + 1$, L at $t\&t + 1$ and R at $t\&t + 1$. Then, the images and depths are fused in a central virtual 360° image. This is useful to increase the reconstructible area around the car and replace inaccurate depths near epipoles of a pair by more accurate depths in another pair. Lastly, the virtual image is segmented into planes using assumptions about the scene: one horizontal plane for the ground, several vertical planes for walls and facades. This is useful to smooth the reconstruction and remove outliers. A set of hypothesis planes is computed by Hough voting in the virtual image, then a plane is selected for each superpixel of the virtual image by minimizing a discrete minimization problem (which has similarities to that in section 3.1.5).

3.3.2. *Cylindrical rectifications*

Both spherical (Li 2006) and cylindrical rectifications are discussed in detail in Bartczak et al. (2007). In both cases, the epipolar plane (which includes a point $\mathbf{P}$ and the two camera centers $\mathbf{C}$ and $\mathbf{C}'$) is parameterized by an angle $t \in [0, 2\pi]$. Latitude $p \in [0, \pi]$ in the spherical case is replaced by height $h \in \mathbb{R}$ in the cylindrical case for a given cylinder radius (Figure 3.5(b)). Since many stereo methods assign integer disparities to pixels in rectified images, it is interesting to explain how this 2D discretization propagates to a 3D discretization that approximates the scene. The disparities are $p - p'$ in the spherical case and $h - h'$ in the cylindrical case, up to the sampling scale in the images. In the former, the iso-disparity surfaces are tori centered at the line segment $\mathbf{CC}'$. In the latter, these surfaces are cylinders with the same axis $(\mathbf{CC}')$. Their densities in 3D increase with the disparity modulus.

In Kang and Szeliski (1997), cylindrical images are computed at several view-points by stitching several images taken by a camera that rotates around a vertical axis (as if a virtual multi-camera takes the original images). An advantage is a high resolution of the cylinders in spite of the low resolution cameras at this time. Once the relative poses of the cylinders are estimated, a basic (WTA) multi-baseline stereo is applied: back-project each pixel of a reference cylinder with hypothesis depths, estimate a photo-consistency in local windows by projecting the obtained 3D point to all cylinders, then select the depth with the best photo-consistency for each pixel. Here, the cylinders have vertical axes and the epipolar curves (where project the 3D points) are sinusoids.

A stereo pair formed by two catadioptric cameras is introduced in Gluckman et al. (1998). Each catadioptric camera is central and composed of a parabolic mirror and an orthographic projection camera. Since they have the same (vertical) symmetry axis, the epipolar curves are simple: radial lines. The radial epipolar lines become parallel after projection of the images in cylindrical images, then a standard window-based correlation (WTA) is applied in the cylinders. This is a first step toward real-time omnidirectional stereo: 7 fps for 600×60 pixels with 32 hypothesis depths.

Cylindrical images are computed by Bunschoten and Krose (2003) at several view-points by cylindrical projections of images taken by a central catadioptric camera (with a hyperbolic mirror) mounted on a mobile robot, after the computation of the camera poses. Then, a simple multi-baseline stereo is applied, which is similar to that in Kang and Szeliski (1997) (neither disparity gradient limit nor ordering). The axes of the virtual cylinders follow the same direction (vertical); this accelerates the mappings between cylinders involved in the photo-consistency computation (sum of squared differences in 11×11 windows). The inaccuracy of the reconstruction is partly due to a low resolution: the cylindrical images are 720×120 (i.e. 2 pixels per degree in the horizontal plane) and 25 hypothesis depths per pixel of the reference cylinder.

A work by Gonzales-Barbosa and Lacroix (2005) deals with a catadioptric camera composed of a perspective camera and two mirrors (parabolic and spherical), where rays are successively reflected. This system is slightly non-central due to misalignment between the mirrors and camera, and is approximated by a virtual central camera. Because of this approximation, the rectification of two input images becomes possible with parallel epipolar lines (rows in images). This implies that the axis of the virtual cylinders, where the images are projected, is parallel to the baseline between the two view points. Then a standard (unspecified) stereo method is applied.

The catadioptric stereo system in He et al. (2007) is mounted on a mobile robot and composed of two hyperbolic mirrors in front of a perspective camera that are coaxial. The mirror that is closest to the camera is projected into an annulus; it has a hole at its center to project the other mirror into a disc surrounded by the annulus. Because of this configuration, a scene point can have two distinct images (only one mirror reflection per ray) and a simple epipolar constraint: the same radial epipolar line. The baseline, which is roughly the distance between the two mirrors, is about 20 cm. Both the disc and annulus parts of the catadioptric image are warped to cylinder images, such that each pair of corresponding epipolar lines becomes a same column. Then a graph-cut method with a second-order prior smoothness is applied on the two cylinder images (section 3.1.5).

3.3.3. *Planar rectifications*

In an early work (Fleck et al. 2005), several images are taken by a catadioptric camera mounted on a mobile robot, the geometry is estimated (using a laser scanner) and a depth map is computed for each triple of consecutive images. Each pixel of the reference image (of a triple) is matched to 50 candidate points in the previous and next images along epipolar curves to define the photo-consistency volume. These computations are accelerated by local planar rectifications of the search regions. Then, the depth map is computed using a graph-cut method (section 3.1.5) with a first-order prior smoothness and by including an additional cost for each occluded (unmatched) pixel. A post-processing greatly improves the results (sub-disparity refinement, removal of epipolar neighborhood, floor correction, filling of small gaps).

Methods designed for perspective images are applied in Jancosek (2014) to panoramic images of Google Street View after image conversions using planar rectifications. This is not optimal due to the central assumption (section 3.2.1), nevertheless large-scale 3D models are obtained like this. Experimented methods include an improved patch-propagation (section 3.1.6) and an improved graph-cut method that estimates a surface as a sub-complex of a 3D Delaunay triangulation (section 3.1.7). The latter improves reconstructed thin structures by favoring inside tetrahedra where the visibility gradient is high.

A work by Huang et al. (2017) reconstructs a 360° video so that it can be replayed in VR headsets with 6 DoF, that is, allowing not only rotations, but also small translations of the user's head. First, an SfM provides the camera poses and a sparse cloud of 3D points. Then an accelerated version (in Shen (2013)) of patch propagation (section 3.1.6) is adapted for the input spherical images. For each image, a second image in the 360° video is selected to obtain a stereo pair with a "good" baseline, which is not too large to allow matching (e.g. good common coverage) and not too small to allow reconstruction (section 3.1.2). The selection is based on angles between reconstruction rays of the same point in the sparse cloud (not too large and not too small angles). Local planar rectifications of the spherical images are used to compute the correlation score between matched points. A Samsung Gear 360 camera and an Oculus Rift are used in the experiments.

3.3.4. *Sphere sweeping*

The method in Im et al. (2016) deals with small baseline. It takes a short video of a bi-camera (a Ricoh Theta S in the experiments) as input, then estimates the geometry of the image sequence, and finally generates a depth map in a reference image by adapting the plane sweeping method (section 3.1.4). Here, each input image concatenates two original fisheye images taken in two opposite directions. This is better for accuracy than stitched images. For each of the two reference sub-images, the sweeping plane is replaced by a sweeping half-sphere centered on a camera center (with a discrete set of radii) to deal with the large field of view. For each point in a half-sphere obtained by back-projecting a pixel in the reference image, a photo-consistency (luminance variance) is computed by projecting the point in all images (Figure 3.6). Then the depth map is improved by a method based on non-local aggregation in the cost volume, which is better than standard local aggregation for the low textured image regions.

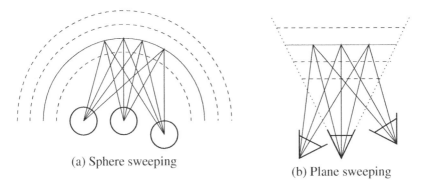

(a) Sphere sweeping

(b) Plane sweeping

Figure 3.6. *Sphere/plane sweeping. The bold spheres are input images. The virtual (half-)spheres, which include 3D points projected to images for photo-consistency computations, are centered on the reference image sphere. Only the current virtual sphere has a continuous pencil stroke. Plane sweeping for perspective cameras is also shown for comparison*

In Won et al. (2019), another method named SweepNet deals with large baseline: four fisheye cameras (220° FoV) near the four corners of the roof of a minivan and pointing to four orthogonal directions. The set of sweeping spheres has 192 spheres centered at the middle of the capture system, such that their inverse radii form an arithmetic progression. For each sphere, the four images are back-projected to the sphere and a matching cost is computed by a method based on deep learning. The neural network takes two 600×150 spherical images as input (one per back-projected image with the same radius, polar regions are ignored) and predicts a cost in range $[0, 1]$ for each pixel, which is 0 if the pixel depth is equal to the sphere radius. It includes 20 convolution layers, five fully connected layers and a final sigmoid layer. The training is done with a synthetic urban dataset and a binary cross-entropy loss between ground truth Boolean (0 = good depth, 1 = bad depth) and the predicted cost. Then, the cost volume is obtained by averaging the cost for image pairs and concatenating for each radius. Lastly, a depth map is computed using SGM (section 3.1.3). Due to the network, which uses the global context of spherical images, the cost volume is cleaner (has less false positives) than those obtained by previous methods, for example, by using zero-mean normalized cross-correlation.

3.3.5. *Neither sweeping nor standard rectification*

Both scene surface and relative pose from two images taken by a catadioptric camera are simultaneously estimated in Pretto et al. (2011). The surface is defined by a 2D Delaunay triangulation in an image whose vertices are back-projected using depth parameters. Some of the vertices are located at point and edge features for a better (piece-wise planar) scene approximation. Since the camera is mounted on the roof of a car, the camera motion is restricted to be planar with an instantaneous center of rotation. The method estimates the pose parameters and the inverse depths of vertices that minimize the sum over the images of the least squares of differences of luminances at the projections of the surface in the two images. The occluded areas are neglected since the two images are successive. The minimization is based on a gradient descent, that is initialized with depth inverse equal to zero (infinite surface) and no camera motion. In the experiments, the surface is deformed along a 1.4 km long trajectory forming a closed loop using 1,800 1,032 × 778 images (the drift is 40 m).

The approach in Pathak et al. (2016) takes two spherical images with small baseline and arbitrary relative rotation (taken by a Ricoh Theta S) as input. A first estimate of the geometry is done using only a matching of sparse image features. It is approximate, but it allows to correct the relative rotation between the two spherical images, that is, one of the two images is corrected by rotating its spherical coordinate system (the color of a pixel in the rotated image is defined as follows: compute its ray direction by back-projection, multiply it by the rotation, get the color that projects the direction). Then, an optical flow method is applied between two spherical images in a more favorable experimental context where their relative motion is a small translation. The optical flow is based on deep learning and trained by using standard (non-omnidirectional) images. Lastly, the geometry is refined by using a greater number of correspondences and a 3D point cloud is reconstructed from the correspondences.

A method is proposed to compute a depth map between images taken by two general central cameras (Roxas and Oishi 2020). First, one image is pre-rotated such that the residual motion between the two images is a pure translation (as in Pathak et al. (2016)). As a result, pairs of corresponding epipolar curves become the same curves in the two images (this is a kind of rectification with smaller distortions near the epipoles and without the constraint that the epipolar curves are lines). Then, a 1D disparity can be defined as the signed arc length between two points in such a curve. A pre-processing (curve tangent at every pixel) accelerates the computation of one of the two points from the other point and a disparity value. Lastly, this fast computation is used to estimate the disparities for all pixels in the reference image that minimize a global cost function with a data term and a smoothing term (section 3.1.5). The former is the absolute difference of luminances. The latter is an anisotropic smoothing that prefers depth discontinuities along image edges. An adequate minimization method is required (primal-dual algorithm) in a hierarchical scheme for large disparities. The approach runs at 10 fps for two 848×800 videos taken by a rigid pair of fisheye cameras.

3.4. Reconstruction from only one central image

The methods summarized in section 3.4 reconstruct an environment by taking only one image provided by an omnidirectional camera as input. We assume that the image calibration is known and that this image is obtained by stitching if a multi-camera is used. Since reconstructing from one view is more difficult than reconstructing from several views, strong prior knowledge or scene constraints are required in addition to the input image. This is one reason why these environments have only been, up until now, indoor environments. Indeed, indoor environments have strong priors: horizontal floor and ceiling, vertical walls that are orthogonal most of the time, doors and windows with standard heights and widths whose boundaries are horizontal or vertical.

There are two kinds of methods: those that explicitly use such geometric constraints (see section 3.4.1) and those that are based on deep learning (see section 3.4.2). The latter are supervised learning methods since they require datasets for training that collect omnidirectional images and their depth maps. The datasets can be obtained from rendering of synthetic scenes or from real scenes acquired by RGB-D cameras (i.e. Chang et al. (2017)).

3.4.1. *Explicit use of geometric constraints*

In an early work (Sturm 2000), a user manually provides all constraints and selects points in the image to be reconstructed. First, parallelism and perpendicularity constraints are used to estimate directions in the 3D coordinates system of the omnidirectional camera. The direction of parallel lines are estimated by SVD once a pair of points are clicked for each line in the image (i.e. vertical lines at the intersection of two walls). A plane normal is known if the plane is orthogonal to a known line direction or if it is parallel to several lines with known directions. Then, the points selected in the image can be progressively reconstructed up to a global scale factor defined by the depth of an initial point. If we know a point in a line whose direction (or in a plane whose normal) is known, then we know all points in the line (or plane) from their projections, including points that are shared by other lines (or planes) whose directions are known, and so on. Lastly, the user draws the boundaries of the reconstructed planes in the image and obtains a simplified 3D model. Details and experiments are given for a catadioptric camera formed by a parabolic mirror and an orthographic camera.

An automatic method (Pintore et al. 2016) assumes that the scene has vertical walls, horizontal ceiling and floor, and the observer's height is known. The input is a spherical image whose zenith axis is vertical (every image column is back-projected as a vertical half-plane). First, the image is segmented in the floor, ceiling and walls by using superpixel segmentation, texture homogeneity and vertical convexity in the image. For each x, there are y_c and y_f, such that pixel (x, y) is in the ceiling or wall or floor if $y \in [0, y_c]$ or $y \in [y_c, y_f]$ or $y \in [y_f, y_{max}]$, respectively. The pixels $(x, y_c(x))$ are back-projected on a horizontal plane at height h_c. The resulting 3D points form an approximate boundary of the ceiling if h_c is the height of the ceiling. Similarly, $(x, y_f(x))$ are back-projected at height $-h_f$ and we obtain an approximation of the floor boundary if $-h_f$ is the observer height. Then, the method estimates the height of the ceiling such that the two boundaries are similar after their orthogonal projections in a horizontal plane. It also estimates a common boundary projection in the horizontal plane by using a Canny edge map. Lastly, a 3D simplified room is obtained as a right cylinder whose base is the polygonized boundary and with the vertical direction. Several reconstructed rooms of a level can be registered into a single 3D model because of their relative poses estimated by an IMU of a standard mobile device (which also takes one spherical image per room).

3.4.2. *Deep learning*

The first deep learning approach that predicts a depth map from a spherical image is proposed in Zioulis et al. (2018). First, a dataset composed of 35×10^3 360° images with their depth maps (which are 512×256) is generated from several previous datasets that include both synthetic and real indoor scenes acquired by RGB-D cameras. These images are vertically aligned and do not have stitching artifacts. Then, two fully convolutional encoder-decoder architectures are proposed. One of them (RectNet) deals with the high distortions near the poles by increasing the receptive field of neurons using dilated convolutions. The training loss is a sum, for several scales, of the L2 loss between the predicted and ground-truth depth maps, and the squared moduli of depth gradients. The incomplete pixels (images and depths can have holes) are discarded in the sum to help the training. In the experiments, the previous methods (which are trained with perspective images) provide worse results, whether they are applied directly on 360° images or on individual faces of cubes obtained from 360° images.

Another approach (Eder et al. 2019) improves the previous one by enforcing piece-wise planar constraints to reconstruct indoor environments. The neural network not only predicts a depth map, but also a surface normal map and a curvature map: it has an encoder-decoder architecture with two decoder branches, the first for depths and the second for normals and curvatures. It takes a spherical image with supplementary maps of longitude and latitude pixel coordinates, which help to distinguish the image areas with different magnitudes of distortion, as input. Then, a previous dataset is augmented by the normals and curvatures (computed from depths) and the training loss is a sum of terms to enforce piece-wise planar constraints, including a L2 loss between the predicted and ground-truth depth maps, a L2 loss between the predicted and ground-truth curvature, a L1 norm of curvature (it should be 0 except at plane boundaries) and a loss for consistency between predicted normals and predicted depths. Lastly, a 3D model is obtained by meshing an improved version of the output depth map (with sharpened planar boundaries) because of the normals and curvatures.

The input of the BiFuse neural network (Wang et al. 2020) is two 360° images taken at the same viewpoint, but with complementary projections: equirectangular (e) and cubemap (c). The e-projection mimics the peripheral vision of the human eye thanks to its large FoV, but has large distortions. The c-projection mimics the foveal vision without large distortion, but has discontinuities at the cube edges. The network is composed of two encoder-decoder branches, a e-branch and a c-branch, each predicts a depth map from a 360° image with the same projection. The two branches are connected at several same levels by fusion blocks for sharing advantages; each fusion block estimates a mask that decides how to combine the features of the level for the next level. The network is ended by a convolution block that combines the e-depths and the c-depths into a final equirectangular depth map. The training has three steps: first train the e-branch and the c-branch independently, then, only train the fusion blocks, lastly, train the complete network. Each step uses a L2 loss between predicted and ground truth depth maps. The method compares favorably to previous work, for example, sharper depths at wall boundaries.

3.5. Reconstruction using stationary non-central camera

A 360° professional VR video capture system and a generic software for depth reconstruction/image rendering are presented in Pozo et al. (2019). The capture system is a spherical rig of 16 fisheye cameras with 1 m diameter (Figure 3.7(a)), which has good overlap for reconstruction. Most 3D points are visible in six to seven cameras. For each frame of each camera, a depth map is reconstructed by using a propagation method, which has similarities with those in section 3.1.6. It includes several ingredients: image rectifications, search for 1D disparities between the reference image and many others by minimizing the zero-mean sum of squared differences over small windows in a hierarchical scheme, pixel-wise propagation in a 5×5 neighborhood, spatial filtering to smooth the disparity and preserve the edges, temporal filtering (with previous and next frames) to avoid depth flickering. If the capture system does not move, foreground masks are also used to accelerate the depth computation. They are provided by a standard background subtraction and restrict the depth computations in the detected foreground (i.e. the moving objects). The rendering step is real-time thanks to GPU, and combines the original images and their depth maps in 360° images with several formats: cubemap and equirectangular projections, left-right eye renders for VR headsets.

Multi-depth panoramas (MDP) are introduced in Lin et al. (2020) for local view synthesis from images taken by a ring of cameras (a 360° multi-camera like the Yi-Halo in Figure 3.7(b), or the GoPro Odyssey in Figure 3.7(c). MDP are several coaxial RGBDA panoramas: each cylindrical image has both depth and alpha channel maps. Many panoramas are better than a single one to render parts of the scene that are disoccluded due to translations of the viewer. They are computed as follows. First, a reduced version of MDP is computed for each reference image and four neighboring images in the ring by using a deep learning method designed for standard (non-omnidirectional) images. A reduced version is composed of several fronto-parallel planes (for the reference image) with RGBA images. The neural network for learning is not 2D, but 3D and takes a plane sweep volume as input, that is, a set of images obtained by warping the input images to parallel planes with several hypothesis depths (their inverses form an arithmetic progression). It is trained from photo-realistic images synthesized by 21 large-scale (indoor and outdoor) scenes. Then, the reduced versions of MDP are merged to MDP. The 3D space is divided in a few coaxial cylindrical regions (about 5), and in each of them, the parts of the reduced versions are merged to one RGBDA panorama using cylindrical coordinates and alpha compositing. Lastly, a novel image is synthesized from MDP by projecting the panoramas as point clouds in a back-to-front order using alpha compositing and z-buffering.

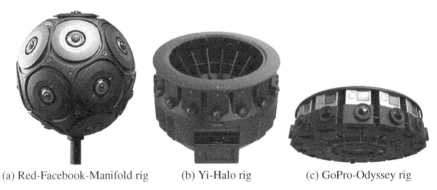

(a) Red-Facebook-Manifold rig   (b) Yi-Halo rig   (c) GoPro-Odyssey rig

Figure 3.7. *Stationary non-central cameras for reconstruction from videos. Every rig has 16 cameras mounted in a spherical or a cylindrical scheme. For a color version of this figure, see www.iste.co.uk/vasseur/omnidirectional.zip*

3.6. Reconstruction by a moving camera

This section surveys works that do not focus on the dense stereo step from a few images (this is done in section 3.3), but on other steps of the reconstruction. The methods in section 3.6.1 mostly merge several local 3D models (e.g. a depth map in a reference image, or a mesh approximation of it) to a global 3D model of the scene. Sections 3.6.2 and 3.6.3 summarize methods that reconstruct a surface from a cloud of 3D points reconstructed from sparse features (points and/or edge curves) detected and matched in the images. Such sparse approaches are interesting for several reasons: to obtain compact models of large-scale environments, to avoid computationally expensive dense stereo (e.g. for applications that do not need a high level of detail), or to initialize dense stereo in other cases. Lastly, section 3.6.4 provides a list of commercial and open source software packages. All these works reconstruct a rigid environment from a set of omnidirectional images taken by a moving camera.

3.6.1. *From local to global models*

3D models of outdoor environments are calculated from hundreds of still images taken by a catadioptric camera (Figure 3.8(b)) mounted on a monopod in Lhuillier (2011). Once the geometry of the image sequence is computed by SfM including self-calibration (Lhuillier 2008) (Figure 3.8(a)), a local 3D model is computed for each image taken as reference (Figure 3.8(c)), then the local models are combined to a global 3D model (Figure 3.8(d)). A local model is computed using planar rectifications (to reduce image distortions) and a pair-wise stereo method as follows. Two catadioptric images are projected to two virtual cubes such that the epipolar lines are parallel except for the faces that contain the epipoles (at the face centers). The match propagation (section 3.1.6) is applied to each of the six pairs of cube faces that are parallel. Once matches are obtained like this for several images pairs that include the reference image, a cloud of 3D points is estimated by ray-intersection of the matches. Furthermore, the reference (catadioptric) image is over-segmented by a 2D mesh that has the following properties: its triangles back-project to similar solid angles and should have low image gradient in their interior, some of its edges approximate the strongest image contours. These triangles are back-projected to approximate the 3D point cloud. This involves several operations, including weighted least squares to fit a plane from 3D points, connections between adjacent triangles and hole filling. Now, the representation of the environment by the set of the local models is highly redundant since there is one omnidirectional local model for each input image. Since scene points are reconstructed several times at different accuracies, the triangles with the worst accuracies are removed. These operations are based on a generic covariance (it does not depend on the camera model), which is related to the standard covariance of the ray-intersection problem.

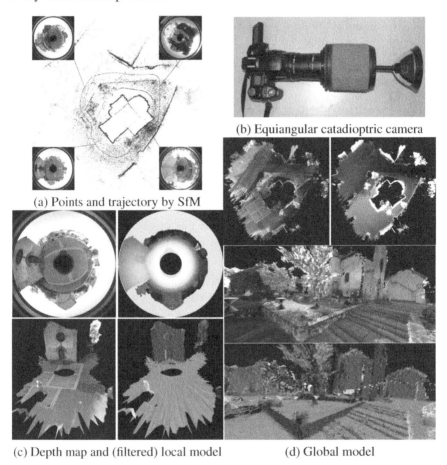

(b) Equiangular catadioptric camera

(a) Points and trajectory by SfM

(c) Depth map and (filtered) local model   (d) Global model

Figure 3.8. *From local to global models (Lhuillier 2011). Note that 208 1632 × 1224 images are taken all around a church using a catadioptric camera and reconstructed by SfM. Each local model (one per image) is a list of triangles approximating a depth map. Their union forms a global model. For a color version of this figure, see www.iste.co.uk/vasseur/omnidirectional.zip*

In Meilland et al. (2015), a graph of RGBD spherical panoramas (with saliency) is computed from videos taken by a multi-camera system mounted on the roof of a car. Each graph edge has a 6 DoF transform that is the relative pose between the two panoramas at its two vertices. These transforms and the selection of multi-camera poses, where the panoramas are computed, are provided by a variant of SfM. The multi-camera system is composed of two triplets of fisheye cameras with a vertical baseline (three on the top and three on the bottom). Each triplet is rectified to a spherical panorama using central approximation such that the epipolar lines are same vertical lines. A dense stereo method like SGM (section 3.1.3) provides the depth maps. In the experiment, the graph representation is estimated for a 1.5 km long trajectory. Note that 310 RGBD panoramas are computed from 7,364*6 images. Then the graph representation is applied to two tasks: real-time localization and view synthesis. The former can estimate the pose of every common camera (monocular, stereo, and omnidirectional) from one image. In the latter, the RGBD sphere whose location is the closest to the virtual viewpoint is used to generate the virtual view.

In Kim and Hilton (2015), both urban and indoor scenes are reconstructed and simplified using cuboids and the Manhattan assumption, that is, by assuming that the scene is piecewise-planar with only three orthogonal directions for the plane normals. First, spherical images with depth maps are computed as in Kim and Hilton (2013) (section 3.3.1). Second, each spherical image is projected to an unit cube whose edges have the three orthogonal directions. This implies that the image edges that correspond to edges between planar pieces of the Manhattan scene have only three configurations in the cube faces: horizontal direction, vertical direction or in a line that crosses the center of the cube face. Thus, the method estimates the rotation, between spherical and cube poses, that maximizes the image edges in these three configurations. Third, the cube faces are segmented into planar regions whose boundaries are preferred in the three configurations. Fourth, planar regions with rectangular boundaries are reconstructed from the segmentation and the depth map by least squares and followed by several refinements. Fifth, the reconstructions are registered in a same 3D coordinate system. Because of the second step, only relative translations between the cube poses need to be estimated. This is done by matching SURF features between images and estimating the translations between the corresponding 3D points. Lastly, the scene is estimated by an union of cuboids, each of them tries to complete a pair of connected and orthogonal rectangles.

A method (Pintore et al. 2018) reconstructs a 3D model of rooms on the same level as several spherical images (taken by a Ricoh Theta S in the experiments). First, a simplified room is reconstructed for each image by using a method similar to that in Pintore et al. (2016) (see section 3.4.1). Second, the simplified rooms are registered in a same 3D coordinate system by SfM. Each of them is a cylinder whose base is a horizontal polygon and axis is the vertical direction. Furthermore, they have the same horizontal plane as floor, and similarly for ceiling. Since the room may require more than one view, a third step merges the simplified rooms that share at least 20% of their bases. Lastly, the user can interactively detect and localize room components mounted on the walls/ceiling/floor (e.g. electric outlets, light switches and air-vents) by 2D search using sliding windows in the back-projected texture of the rooms.

3.6.2. *Sparse approaches for local models*

In Kawanishi et al. (2011), an SfM tracks and reconstructs points and lines in a series of catadioptric images taken in a piece-wise planar environment. Lines are useful for low textured scenes, which have a small number of reliable tracked interest points. Then, a 2D Delaunay triangulation is computed in a reference image whose vertices are points with 3D information, that is, a reconstructed point or a point in a reconstructed line. The textured 3D model of the scene is initialized as the back-projected triangulation using the texture of the reference image. However, the Delaunay edges can be inconsistent with the scene, that is, if a Delaunay edge crosses the projection of a scene edge separating two planar regions of the scene with different colors. This provides visual artifacts when the 3D model is seen from a view point that is not that of the reference image. Such artifacts are removed by a method based on edge flipping of the 2D triangulation (Nakatsuji et al. 2005). This method symmetrically uses two images of the input sequence (including the reference image) and does not need threshold tuning. In short, it tries to flip edges to minimize an image error, that is, it replaces two adjacent triangles **bca** and **bcd** by two other adjacent triangles **adb** and **adc**, such that a discrepancy between a first image and the reprojection of the other image on the first image decreases.

A coarse 3D model of an environment is reconstructed by using a smartphone (Pan et al. 2011) in three steps: first compute a few panoramic images by rotating the smartphone during video acquisition, then robustly match and reconstruct feature points between the panoramic images, last reconstruct a surface by using a 3D Delaunay triangulation (section 3.1.7). The first step uses the central assumption for the generation of a panorama (without tripod), since the distance between scene and camera is quite a bit larger than the distances between different camera centers. Each panorama is done in a few tens of seconds, both rotation estimation and image stitching are in real time (30 Hz). The second step pre-rotates the panoramas around their axis (roughly vertical) to reduce the search space for matching and avoid false matches due to repetitive textures. SfM starts by detecting features points in each panorama and ends by a bundle adjustment for estimating both panorama poses and 3D points. The last step computes the 3D Delaunay triangulation of the reconstructed points. Then, the list of outside tetrahedra grows in the list of all Delaunay tetrahedra (the Delaunay tetrahedra are "carved"). The first outside tetrahedra are finite and include the camera centers (since the camera centers are in the convex hull of the reconstructed points in the omnidirectional context). Lastly, neighboring tetrahedra are progressively added to the set of outside tetrahedra if their intersections by rays are strong enough in a probabilistic sense. In practice, $2,048 \times 512$ panoramas are generated from 320×240 videos, the second and third steps take about 15–25 s for three to seven panoramas.

3.6.3. *Sparse approaches for global models*

Simplified 3D models of outdoor environments are computed from catadioptic images (Lhuillier and Yu 2013) by using a manifold method (section 3.1.7) in two cases: still images taken by a hand-held camera mounted on a monopod (Figure 3.8(b)), or a video taken by a helmet-held camera (Figure 3.3(a)). The computation starts with an adequate SfM algorithm (see: Lhuillier (2008) and Mouragnon et al. (2007), respectively). The output of the latter is similar to that of the former: still images automatically selected from the video and a sparse cloud of points reconstructed from tracked Harris corners. The next computations are the same. The 3D Delaunay triangulation is computed from input points (including a ray intersection counter for each tetrahedron) and the set O of outside tetrahedra continuously grows in the free-space, while maintaining the manifold property on the surface ∂O. Since a point has bad accuracy if it is roughly collinear to the camera locations that reconstruct it, a pre-filtering is needed (reject a point if all angles formed by its reconstruction rays are below a threshold). This degenerate case is not negligible since epipoles are always reprojected in omnidirectional images. Then, the growing has several steps. First, the main part of O is obtained by adding tetrahedra one-by-one to O without topology change: ∂O is homeomorphic to a sphere. However, this is insufficient if the camera trajectory has a complete loop around a large obstacle like a building. Indeed, ∂O needs a handle like a torus. Thus, a second step solves this problem by adding several tetrahedra to O at once. Now, spurious handles can occur and a third step removes the largest ones. The third step has a detect-force-and-repair strategy, which is summarized as follows: get the largest edges of triangles in ∂O, for each one try to force in O the free-space tetrahedra that share the edge and grow O such that ∂O becomes manifold again. The two cases are experimented: a surface with 1.5×10^5 triangles from 343 still images around a church, a surface of a city with 4.1×10^5 triangles from 2.5×10^3 images selected in a video taken at a 1.4 km long trajectory.

The manifold method above is improved in several ways and experimented on videos taken by a helmet-held multi-camera (Lhuillier 2018b). The first step is an SfM designed for such a camera system (Nguyen and Lhuillier 2017): with synchronization, self-calibration and keyframe selection, without image stitching. To capture more scene details, the sparse cloud of 3D points is completed by reconstructing both Harris and curve (Canny edge) points that are matched between the selected images using the match propagation (section 3.1.6) and the epipolar constraint. Then, a 3D Delaunay triangulation is computed from pre-filtered points and O grows in the free-space tetrahedra to maximize the visibility consistency. This growth is not only regularized by manifoldness, but also by lowered genus. Remember that the genus is the number of handles of ∂O, and it is provided by the Euler formula. The lower the genus, the simpler the topology of ∂O. The main idea is to apply operations that can increase the genus (they are similar to those in steps two and three of Lhuillier and Yu (2013)) only if they generate visually non-negligible updates of ∂O. Furthermore, the previous repair operation is replaced by a new one that reduces the risk of failure since it is guided by an analysis of non-manifold vertices and edges (the new one is also quite a bit faster than the previous one). Other improvements include an acceleration of manifold tests and better escapes from local extrema. In the experiments, the input is a set of four 1280×960 videos at 100 Hz taken by walking for 16 minutes in an university campus using a do-it-yourself multi-camera: four GoPro Hero3 cameras mounted on a helmet (Figure 3.3(c)). The 3D model has 4.2×10^6 triangles and is reconstructed using 3.4×10^3 keyframes in less than one minute on a laptop (after the computation of the 3D points). Here, a keyframe is a concatenation of four frames, one per GoPro camera, that have frame-accurate synchronization.

The most notable mistakes of the surface reconstruction methods from sparse point cloud based on 3D Delaunay triangulation are falsely labeled free-space tetrahedra. They occur with both graph-cut- and manifold-based methods (section 3.1.7) and are due to bad points or lack of points. They appear in incomplete shapes like 3D boxes (e.g. buildings) with spurious concavities, thin 1D structures (e.g. posts) that are disconnected, thin 2D structures (e.g. traffic signs) with tunnels connecting both sides. Both Lhuillier (2018a) and Lhuillier (2019) improve these methods thanks to a pre-processing or a post-processing and experiment from input videos taken by a 360° camera. They complete the shapes, that is, relabel matter falsely labeled tetrahedra, by using a prior: the predominance of vertical structures in usual environments. This means that the relabeling from free-space to matter is more important in the vertical direction than in the horizontal ones. Thin 1D structures are completed (Lhuillier 2018a) in a pre-processing by using reconstructed curves from the image gradient. First, structures are detected in the 3D Delaunay triangulation as connected sets of vertices coming from vertical curves, that have small matter neighborhood in the horizontal directions. Then, series of tetrahedra are forced to matter along vertices of connected sets. Lastly, graph-cut and manifold methods can be applied with slight modifications. Several completions in Lhuillier (2019) deal with the 1D/2D/3D structures above as pre-processing of the manifold methods or post-processing of the graph-cut methods. In short, they increase the local-convexity of the shapes: they select a lot of sets of free-space tetrahedra, then all tetrahedra in a set are relabeled matter if there are enough matter tetrahedra that surround the set. Several criteria are used to select the sets and quantify their surrounding matter, most of them encourage the matter completion in the vertical direction. In the experiments from the papers, the input videos are two 2496×2496 videos at 30 Hz provided by a Garmin Virb 360 camera with two opposite fisheye pointing on the left and the right (Figure 3.4). City and town segments are reconstructed by mounting the 360° camera on a helmet or a car and by walking/biking/driving along trajectories up to a few kilometers.

Textured 3D models, that are computed by methods in Lhuillier (2018b, 2019), are also available for interactive exploration using common consumer-grade VR headsets (PC VR[1], Oculus Quest and Go[2], SteamVR workshop[3]) or without VR[4]. One example is detailed in Figure 3.9 for a semi-medieval city reconstructed by a helmet-held GoPro Max 360 camera.

The methods previously summarized in section 3.6.3 are batch methods: they compute a surface once the complete image sequence and its geometry are available. They cannot be used for online applications that need an updated surface during the video acquisition. For this reason, incremental versions of batch methods are developed by taking advantage of both incremental 3D Delaunay triangulation and incremental SfM. A method by Litvinov and Lhuillier (2013) locally updates the manifold surface ∂O encoded in a 3D Delaunay triangulation using new points reconstructed by an incremental SfM (Mouragnon et al. 2007), and experiments on videos taken by an omnidirectional camera. These new points cannot be directly added to the Delaunay, otherwise ∂O becomes non-manifold, and it is difficult to correct this while maintaining a Delaunay triangulation. Thus, the idea is to shrink O before adding the new points, such that ∂O is still manifold and O does not include a tetrahedron that will be destroyed by the addition of points. The shrinking is nothing but an inverted growing with similar operations. It pushes O out of a bounding ball that contains all destroyed tetrahedra. Such a ball is computed thanks to moderate constraints enforced on the new points (bounded distance to the new camera center) and the Delaunay (bounded tetrahedron diameter thanks to Steiner vertices in a regular grid). Once O shrinks, the new points are added to the Delaunay, the set of free-space tetrahedra is locally updated with their ray counters (mostly by using the new reconstruction rays provided by SfM), and the last O grows with similar operations as in the batch versions. This method is improved in Litvinov and Lhuillier (2014) by accelerating the most time consuming operations: the removal of artifacts, including spurious handles. A previous version (Yu and Lhuillier 2012) of these works replaces the shrinking by handling of several levels of sets O at different times, but its time complexity is not bounded if the SfM closes a loop. In the experiments, Litvinov and

---

1 Available at: https://maximelhuillier.fr.

2 Available at: https://sidequestvr.com/user/330664.

3 Available at: https://steamcommunity.com/profiles/76561198051374313/myworkshopfiles/.

4 Available at: https://sketchfab.com/flymax63.

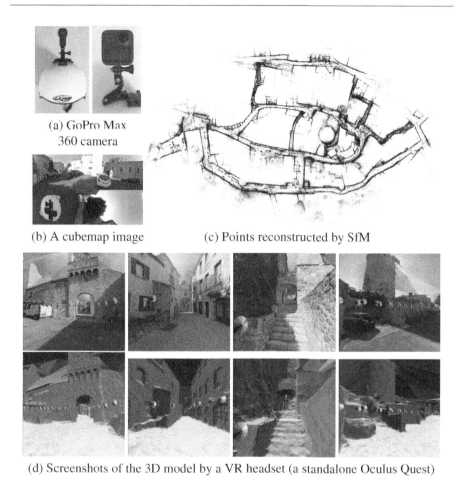

(a) GoPro Max 360 camera

(b) A cubemap image

(c) Points reconstructed by SfM

(d) Screenshots of the 3D model by a VR headset (a standalone Oculus Quest)

Figure 3.9. *3D modeling for VR. A helmet-held GoPro Max 360 camera records two 4,096 × 1,344 videos at 30 Hz by walking 15 minutes in a semi-medieval city. SfM reconstructs 5.9×10^6 points and selects 1.9×10^3 keyframes. The reconstructed surface has 3.7×10^6 triangles. After simplification and texturing, it is explorable like a pedestrian using common consumer-grade VR headsets (https://sidequestvr.com/ app/2277/3d-scan-of-a-semi-medieval-town). For a color version of this figure, see www.iste.co.uk/vasseur/omnidirectional.zip*

Lhuillier (2013, 2014) take six 1024×768 videos at 15 Hz, taken by a $360°$ multi-camera (PointGrey Ladybug in Figure 3.3.(b)) mounted on the roof of a car that moves on a 2.5 km long trajectory in a city, as input. The input video in Yu and Lhuillier (2012) is taken by a hand-held equiangular catadioptric camera (a mirror in front of the Canon Legria HSF10 in Figure 3.3(a)) and by walking on a 800 m long trajectory in a city.

3.6.4. *Available software*

This section provides a list of commercial and open source software packages that reconstruct environment from omnidirectional images in 2020. The list can become non-exhaustive with time. It is restricted to software that compute a dense cloud of 3D points or a triangulated surface. It does not include software that only computes SfM. (SfM is not the topic of the current chapter.) If the omnidirectional camera is a multi-camera, most of the software packages assume that the input images are stitched. Thus, they implicitly assume that both central approximation and calibration are accurate enough for stitching and the downstream 3D computations. A minority of them can take original images of multi-cameras as input and benefit from the assumption of constant relative pose(s) between its monocular component cameras. In this last case, they estimate the relative pose(s) and are the best solution for the geometry accuracy. The catadioptric images could also be used after cylindrical or spherical reprojections. Furthermore, almost all of the listed software packages take still images as input. If the input is a video, the user needs to select images in the video, for example, one image every second by using a tool like FFmpeg (2021). Alternatively, the user can use SfM software designed for multi-cameras to select still images and try to convert the geometry output to the input of a listed software.

The listed software packages are mostly designed for images taken by standard cameras, but have options for omnidirectional images. Metashape (Agisoft 2021) can take equirectangular and cylindrical images as input, but recommend the processing of the original images by a Metashape function instead of stitching by another software. In Barazzetti et al. (2018), both Pix4Dmapper and PhotoScan are experimented with images taken by the Xiaomi Mijia Sphere 360 camera. 3DF Zephyr (*3Dflow* 2018) is one of the only software package that can take a video as input. It also converts equirectangular images into cubemaps for 3D computations, that is, as if six pinhole cameras are used (one per face cube). In Barazzetti et al. (2017), both ContextCapture and PhotoScan are experimented on the Samsung Gear 360 camera. PhotoScan is also experimented in Sun and Zhang (2019) to input 360° videos (taken by Xiaomi Mi Sphere) sampled at different rates and the results are compared to ground truth. There are also open source software packages: MicMac (Rupnik et al. 2017) and Meshroom (AliceVision 2021). The former accepts a lot of camera models including spherical cameras. Furthermore, it can take original images of a multi-camera as input if their relative poses are rotations (estimated by bundle adjustment). It includes SGM (section 3.1.3) among other dense stereo methods. The latter accepts the original images of a multi-camera as input. It includes SGM and a surface reconstruction (Jancosek 2014) that improves a method based on graph-cut and 3D Delaunay triangulation (section 3.1.7).

3.7. Conclusion

This chapter surveys the previous work on the reconstruction of environments from omnidirectional images, in the form of a dense cloud of points or a set of triangles (without focusing on the geometry estimation step). The publication rate in this topic accelerates between 2015-2020 for several reasons, including the apparition on the market of a lot of consumer-grade 360° cameras. Since the experimental context varies a lot, the methods are classified by their assumptions and input: a few images for dense stereo, one central image with strong (indoor) scene prior, videos taken by a stationary non-central camera (the environment can be non-rigid here), and a lot of still or video images taken by a moving camera. Most of them are adapted from popular methods for perspective cameras, which are revisited at the beginning to obtain a more self-contained chapter. The summaries of the methods include the principles, of algorithms, explicit assumptions, target applications and datasets used in experiments.

Thanks to both hardware improvements (resolutions and frame rates increase) and software improvements (e.g. deep learning), we can guess that the consumer-grade 360° cameras will be progressively used to reconstruct large scale and complete environments from terrestrial imagery with minimal effort of the user, in a similar way as the standard cameras for reconstructing limited objects, likes statues or single facade. Up until 2020, very few convincing results have been shown on the internet (e.g. on YouTube and Sketchfab) in spite of available software and potential applications, such as VR. Alternatives to the pure image approach can also appear, such as the use of solid-state LiDAR for robotic applications.

3.8. References

3Dflow Academy (2018). 3Dflow academy – videotutorial 11 – using 360 cameras [Online]. Available at: www.youtube.com/watch?v=SDa2GfFCUg.

Agisoft (2021). Metashape, professional edition [Online]. Available at: www.agisoft.com/downloads/user-manuals.

AliceVision (2021). Meshroom [Online]. Available at: https://alicevision.org/.

Anguelov, D., Dulong, C., Filip, D., Frueh, C., Lafon, S., Lyon, R., Ogale, A., Vincent, L., Weaver, J. (2010). Google Street View: Capturing the world at street level. *IEEE Computer*, 43(6), 32–38.

Arican, Z. and Frossard, P. (2007). Dense disparity estimation from omnidirectional images. *Proc. IEEE Conf. Adv. Video Sign. Surv.*, 399–404.

Barazzetti, L., Previtali, M., Roncoroni, F. (2017). 3D modelling with the Samsung Gear 360. *International Archive of Photogrammetry, Remote Sensing and Spatial Information Sciences*, XLII-2/W3, 85–90.

Barazzetti, L., Previtali, M., Roncoroni, F. (2018). Can we use low-cost 360 degree cameras to create accurate 3D models? *Int. Arch. Photogramm. Remote Sens. Spat. Inf. Sci.*, XLII-2, 69–75.

Bartczak, B., Koser, K., Woelk, F., Koch, R. (2007). Extraction of 3D freeform surfaces as visual landmarks for real-time tracking. *J. Real-Time Im. Pr.*, 2(2), 81–101.

Bunschoten, R. and Krose, B. (2003). Robust scene reconstruction from an omnidirectional vision system. *IEEE Trans. Robot. Autom.*, 19(2), 351–357.

Chang, A., Dai, A., Funkhouser, T., Halber, M., Niessner, M., Savva, M., Song, S., Zeng, A., Zhang, Y. (2017). Matterport3D: Learning from RGB-D data in indoor environments. In *Proc. Int. Conf. 3D Vis.* IEEE, Qingdao.

Eder, M., Moulon, P., Guan, L. (2019). Pano popups: Indoor 3D reconstruction with a plane-aware network. In *Proc. Int. Conf. 3D Vis.* IEEE, Quebec City.

FFmpeg (2021). FFmpeg [Online]. Available at: https://ffmpeg.org.

Fleck, S., Busch, F., Biber, P., Strasser, W., Andreasson, H. (2005). Omnidirectional 3D modeling on a mobile robot using graph cuts. In *Proc. IEEE Int. Conf. Robot. Autom.* Barcelona.

Furukawa, Y. and Hernandez, C. (2015). Multi-view stereo: A tutorial. *Foundations and Trends in Computer Graphics and Vision*, 9(1), 1–148.

Gallup, D., Frahm, J., Mordohai, P., Yang, Q., Pollefeys, M. (2007). Real-time plane-sweeping stereo with multiple sweeping directions. In *Proc. IEEE Conf. Comp. Vis. Pattern Recogn.* Minneapolis.

Gluckman, J., Nayar, S., Thoresz, K. (1998). Real-time omnidirectional and panoramic stereo. *Proc. DARPA Image Understanding Workshop.* Morgan Kaufmann Publishers, Monterey, 1, 299–303.

Gokhool, T., Martins, R., Rives, P., Despres, N. (2015). A compact spherical RGBD keyframe-based representation. *Proc. IEEE Int. Conf. Robot. Automat.* Seattle, 4273–4278.

Gonzales-Barbosa, J. and Lacroix, S. (2005). Fast dense panoramic stereovision. In *Proc. IEEE Int. Conf. Robot. Automat.* Barcelona.

Hartley, R. and Zisserman, A. (2000). *Multiple View Geometry in Computer Vision*, 1st edition. Cambridge University Press.

He, L., Luo, C., Zhu, F., Hao, Y., Ou, J., Zhou, J. (2007). Depth map regeneration via improved graph cuts using a novel omnidirectional stereo sensor. In *Proc. IEEE Int. Conf. Comp. Vis.*, Rio de Janeiro.

Hirschmuller, H. (2005). Accurate and efficient stereo processing by semi-global matching and mutual information. *Proc. IEEE Conf. Comp. Vis. Pattern Recogn.*, San Diego, 2, 807–814.

Huang, J., Chen, Z., Ceylan, D., Jin, H. (2017). 6-DOF VR videos with a single 360-camera. *Proc. IEEE Virtual Reality*, Los Angeles, 37–44.

Im, S., Ha, H., Rameau, F., Jeon, H., Chloe, G., Kweon, I. (2016). All-around depth from small motion with a spherical panoramic camera. In *Proc. Eur. Conf. Comp. Vis.* Springer, Amsterdam.

Jancosek, M. (2014). Large scale surface reconstruction based on point visibility. PhD Thesis, Czech Technical University, Prague.

Kang, S. and Szeliski, R. (1997). 3D scene data recovery using omnidirectional multibaseline stereo. *Int. J. Comput. Vision*, 25(2), 167–183.

Kawanishi, R., Yamashita, A., Kaneko, T. (2011). Three-dimensional environment modeling based on structure from motion with point and line features by using omnidirectional camera. *InTech*, December 14th.

Kim, H. and Hilton, A. (2013). 3D scene reconstruction from multiple spherical stereo pairs. *Int. J. Comput. Vision*, 104(1), 94–116.

Kim, H. and Hilton, A. (2015). Block world reconstruction from spherical stereo image pairs. *Comput. Vis. Image Und.*, 139, 104–121.

Lhuillier, M. (2008). Automatic scene structure and camera motion using a catadioptric system. *Comput. Vis. Image Und.*, 109(2), 186–203.

Lhuillier, M. (2011). A generic error model and its application to automatic 3D modeling of scenes using a catadioptric camera. *Int. J. Comput. Vision*, 91(2), 175–199.

Lhuillier, M. (2018a). Improving thin structures in surface reconstruction from sparse point cloud. In *Eur. Conf. Comp. Vis. Workshop*. Springer, Munich.

Lhuillier, M. (2018b). Surface reconstruction from a sparse point cloud by enforcing visibility consistency and topology constraints. *Comput. Vis. Image Und.*, 175, 52–71.

Lhuillier, M. (2019). Local convexity reinforcement for scene reconstruction from sparse point clouds. In *Proc. Int. Conf. 3D Immer*. IEEE, Brussels.

Lhuillier, M. and Quan, L. (2002). Match propagation for image-based modeling and rendering. *IEEE Trans. Pattern Anal. Mach. Intell.*, 24(8), 1140–1146.

Lhuillier, M. and Yu, S. (2013). Manifold surface reconstruction of an environment from sparse structure-from-motion data. *Comput. Vis. Image Und.*, 117(11), 1628–1644.

Li, S. (2006). Real-time spherical stereo. In *Proc. IEEE Conf. Pattern Recogn.* Hong Kong, 3, 1046–1049.

Lin, K., Xu, Z., Mildenhall, B., Srinivasan, P., Hold-Geoffroy, Y., DiVerdi, S., Sun, Q., Sunkavalli, K., Ramamoorthi, R. (2020). Deep multi depth panoramas for view synthesis. In *Proc. Eur. Conf. Comp. Vis.* Springer, Glasgow.

Litvinov, V. and Lhuillier, M. (2013). Incremental solid modeling from sparse omnidirectional structure-from-motion data. In *Proc. Brit. Mach. Vis. Conf.* BMVA Press, Bristol.

Litvinov, V. and Lhuillier, M. (2014). Incremental solid modeling from sparse structure-from-motion data with improved visual artifacts removal. In *Proc. IEEE Conf. Pattern Recogn.*, Stockholm.

Meilland, M., Comport, A., Rives, P. (2015). Dense omnidirectional RGB-D mapping of large-scale outdoor environments for real-time localization and autonomous navigation. *J. Field Rob.*, 32(4), 474–503.

Mouragnon, E., Lhuillier, M., Dhome, M., Dekeyser, F., Sayd, P. (2007). Generic and real time structure from motion. In *Proc. Brit. Mach. Vis. Conf.* BMVA Press, Warwick.

Nakatsuji, A., Sugaya, Y., Kanatani, K. (2005). Optimizing a triangular mesh for shape reconstruction from images. *IEICE Trans. Information and Systems*, 88(10), 2269–2276.

Nguyen, T.-T. and Lhuillier, M. (2017). Self-calibration of omnidirectional multi-cameras including synchronization and rolling shutter. *Comput. Vis. Image Und.*, 162, 166–184.

Pan, Q., Arth, C., Reitmayr, G., Rosten, E., Drummond, T. (2011). Rapid scene reconstruction on mobile phones from panoramic images. In *Proc. 10th IEEE Int. Symp. Mixed Augm. Real.*, Basel.

Pathak, S., Moro, A., Fujii, H., Yamashita, A., Asama, H. (2016). 3D reconstruction of structures using spherical cameras with small motion. In *Proc. 16th Int. Conf. Contr., Autom. Syst.* IEEE, Gyeongju.

Pintore, G., Garro, V., Ganovelli, F., Agus, M., Gobbetti, E. (2016). Omnidirectional image capture on mobile devices for fast automatic generation of 2.5D indoor maps. In *Proc. IEEE Winter Conf. Appl. Comp. Vis.*, Lake Placid.

Pintore, G., Pintus, R., Ganovelli, F., Scopigno, R., Gobbetti, E. (2018). Recovering 3D existing-conditions of indoor structures from spherical images. *Comput. Graphics*, 77, 16–29.

Pozo, A., Toksvig, M., Schrager, T.F., Hsu, J., Mathur, U., Sorkine-Hornung, A., Szeliski, R., Cabral, B. (2019). An integrated 6DoF video camera and system design. *ACM Trans. Graphics*, 38(6), 1–16.

Pretto, A., Menegatti, E., Pagello, E. (2011). Omnidirectional dense large-scale mapping and navigation based on meaningful triangulation. In *Proc. IEEE Int. Conf. Robot. Automat.*, Shanghai.

Roxas, M. and Oishi, T. (2020). Variational fisheye stereo. *IEEE Robot. Autonom. Lett.*, 5(2), 1303–1310.

Rupnik, E., Daakir, M., Deseilligny, M. (2017). MICMAC – A free, open-source solution for photogrammetry. *Open Geospatial Data, Software and Standards*, 2(1), 14.

Schönbein, M. and Geiger, A. (2014). Omnidirectional 3D reconstruction in augmented Manhattan worlds. In *Proc. IEEE Int. Conf. Robot. Automat.*, Chicago.

Shen, S. (2013). Accurate multiple view 3D reconstruction using patch-based stereo for large-scale scenes. *IEEE Trans. Image Process.*, 22(5), 1901–1914.

Sturm, P. (2000). A method for 3D reconstruction of piecewise planar objects from single panoramic image. In *Proc. IEEE Workshop on Omnidirectional Vision (OMNIVIS).*, Hilton Head.

Sun, Z. and Zhang, Y. (2019). Accuracy evaluation of videogrammetry using a low-cost spherical camera for narrow architectural heritage: An observation study with variable baseline and blur filters. *Sensors*, 19(3), 496.

Thatte, J., Boin, J., Laksham, H., Girod, B. (2016). Depth augmented stereo panorama (DASP) for cinematic virtual reality with head-motion parallax. In *Proc. IEEE Int. Conf. Multim. Expo.*, Seattle.

Vu, H., Labatut, P., Pons, J., Keriven, R. (2011). High accuracy and visibility-consistent dense multiview stereo. *IEEE Trans. Pattern Anal. Mach. Intell.*, 34(5), 889–901.

Wang, F., Yeh, Y., Sun, M., Chiu, W., Tsai, Y. (2020). BiFuse: Monocular 360 depth estimation via bi-projection fusion. In *Proc. IEEE Conf. Comp. Vis. Pattern Recogn.*, 13–19 June.

Wikipedia (n.d.). List of omnidirectional (360-degree) cameras [Online]. Available at: https://en.wikipedia.org/wiki/List_of_omnidirectional_(360-degree)_cameras.

Won, C., Ryu, J., Lim, J. (2019). SweepNet: Wide-baseline omnidirectional depth estimation. In *Proc. IEEE Int. Conf. Robot. Automat.*, Montreal.

Yu, S. and Lhuillier, M. (2012). Incremental reconstruction of manifold surface from sparse visual mapping. In *Proc. 2nd Int. Conf. 3D Imag. Mod. Proc. Visualiz. Transm.* IEEE, Zurich.

Zioulis, N., Karakottas, A., Zarpalas, D., Daras, P. (2018). Omnidepth: Dense depth estimation for indoors spherical panoramas. In *Proc. Eur. Conf. Comp. Vis.* Springer, Munich, 453–471.

4

Catadioptric Processing and Adaptations

Fatima AZIZ[1], Ouiddad LABBANI-IGBIDA[1] and
Cédric DEMONCEAUX[2]

[1] *XLIM Laboratory, University of Limoges, France*
[2] *ImViA Laboratory, University of Burgundy, Dijon, France*

Catadioptric images present strong anamorphoses related to the projective deformations by the surfaces of the projection mirrors. Perspective processing can still be applied in limited situations. This chapter presents adapted processing approaches to catadioptric imaging that take into account these deformations and propose adaptations of the associated processing. The chapter mainly focuses on approaches that apply directly in the original catadioptric image plane without using intermediate reprojection surfaces. Examples of applications (image smoothing, Gaussian difference filtering, edge detection, linear filtering, interest point extraction, scale-space analysis) are presented and illustrate the effect of these approaches in comparison to conventional processing in perspective imaging.

4.1. Introduction

Catadioptric imaging, while offering a wide field of view, comes with a trade-off of high anamorphosis that makes image processing and interpretation complex. The latter is due to the imaging process, where the

3D scene is first projected onto a curved mirror before being captured by a conventional camera. Consequently, the formation of the image in catadioptric vision undergoes a nonlinear transformation due to the reflection on the mirror. This introduces difficulties in modeling the mapping function and in processing image information.

Two main different approaches can be considered for the processing of catadioptric images:

– *Perspective processing*: most approaches neglect the nonlinear anamorphosis of the image and use processing tools borrowed from well-developed and known concepts of perspective imaging, applying for, for example, basic smoothing, primitive segmentation, feature extraction, partial differential operators and so on. These tools indeed provide many processing solutions in catadioptric imaging applications. They are, however, not capable of handling high geometric distortions and non-uniform resolution, limiting their use to simple processing applications where accuracy is not much of an issue.

– *Adapted processing*: few works deal with specific nonlinear processing by nonlinearly adapting the neighborhood of each image point for each kernel-based processing or by totally revisiting the mathematical operators to take into account the transformation introduced by the mirror. These formulations could be performed directly in the 2D image plane or by wrapping the image information on the equivalence sphere.

In this chapter, we will focus on the second approach, which develops specific tools for catadioptric imaging, leveraging on uniform equivalence models or adapting neighborhood metrics and concepts to the distortion inherent in such images.

4.2. Preliminary concepts

Let us first introduce some fundamental and generic concepts of formation and models of catadioptric images, as well as the concepts of Riemannian geometry which will be used to adapt the processing.

4.2.1. *Spherical equivalence models*

The spherical projection model (Geyer and Daniilidis 2000; Figure 4.1) unifies all types of conventional cameras or central catadioptric systems that respect the single viewpoint (SVP) constraint (Baker and Nayar 1999). It is obtained by a two-step projection: firstly, a perspective projection of a scene point $P(X;Y;Z)$ onto the surface of a unit sphere centered on the SVP, followed by a projection from a point belonging to the vertical axis of the sphere toward the catadioptric image plane. The unified spherical model therefore has two parameters, ξ and φ, determined by the shape of the mirror and its eccentricity as described in Table 4.1 and expressed by equation [4.1].

$$x = \frac{\xi - \varphi}{\xi\|P\| - Z}X \qquad y = \frac{\xi - \varphi}{\xi\|P\| - Z}Y \qquad\qquad [4.1]$$

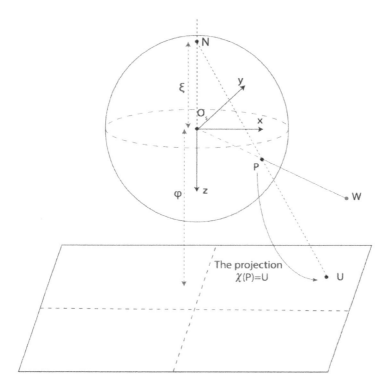

Figure 4.1. *Central catadioptric image formation using the unified projection model as defined in Geyer and Daniilidis (2000). The first mapping produces the intersection point P_s on the sphere, followed by a projection from a specific point on the z-axis, defined by a parameter ξ depending on the mirror shape, to the image point U. The distance between the image plane and the center of the sphere O_s is defined by the parameter φ. For a color version of this figure, see www.iste.co.uk/vasseur/omnidirectional.zip*

Mirror	ξ	φ
Parabolic ($e = 1$)	1	$2p - 1$
Hyperbolic ($e > 1$) & Elliptical ($e < 1$)	$\dfrac{2e}{1 + e^2}$	$\dfrac{2e(2p - 1)}{1 + e^2}$
Planar ($e \to \infty$)	0	f

Table 4.1. *Parameters ξ and φ of the equivalence sphere model. e is the eccentricity of the conic section defined by the reflecting surface, f is the focal length and $4p$ is the latus rectum, that is, the length of the line segment created by the two points of intersection of the conic section with the line passing through the mirror focus (see Figure 4.1)*

One major interest is to wrap the catadioptric image on a uniform manifold where one can call upon spherical concepts and calculus on the spherical manifold or its tangent spaces. We refer the reader to Chapters 1 and 2 for detailed developments of catadioptric geometry and models.

4.2.2. *Differential calculus and Riemannian geometry*

This section introduces some basic concepts of differential geometry: manifolds, their derivatives and tangent spaces. The definition of a Riemannian metric enables length and angle measurements on tangent spaces giving rise to the notions of curve length, geodesics and thereby the basic elements for embedding catadioptric adapted processing. For an in-depth study of Riemannian geometry, we refer the reader to Lee (2006) and Petersen (2006).

DEFINITION 4.1 (Manifold).– *A **manifold** M is a metric space with the following property: each point, $\mathbf{p} \in M$, has a local neighborhood that is homeomorphic to $\mathbb{R}^n$ (where n is the dimension of the space).*

The Euclidean plane $\mathbb{R}^2$ and the sphere $\mathbb{S}^2$ are examples of two-dimensional manifolds also called surfaces. Intuitively, a manifold is a collection of points that locally, but not globally, resembles Euclidean space.

A smooth differential manifold may be endowed with an additional local structure called formally a "Riemannian metric".

DEFINITION 4.2 (Riemannian metric).– *A **Riemannian metric** on a smooth manifold M is a 2-tensor field g that is symmetric (i.e. $g(\mathbf{x}, \mathbf{y}) = g(\mathbf{y}, \mathbf{x})$) and positive definite (i.e. $g(\mathbf{x}, \mathbf{x}) > 0$ if $\mathbf{x} \neq 0$). A Riemannian metric thus* *determines an inner product on each tangent space T_pM, which is typically written as:$\langle \mathbf{x}, \mathbf{y} \rangle = g(\mathbf{x}, \mathbf{y})$ for $\mathbf{x}, \mathbf{y} \in T_pM$. A manifold together with a given Riemannian metric defines a Riemannian manifold.*

For instance, on $\mathbb{R}^n$, the Riemannian metric is given by the standard inner product $g(\mathbf{x}, \mathbf{y}) = \langle \mathbf{x}, \mathbf{y} \rangle = \mathbf{x} \cdot \mathbf{y}$ for all $\mathbf{x}, \mathbf{y} \in T_p\mathbb{R}^n$, for all $\mathbf{p} \in \mathbb{R}^n$. We call $\mathbb{R}^n$ with this Riemannian metric, *Euclidean space* formed by the usual Euclidean dot product $(\mathbb{R}^n, \langle \cdot, \cdot \rangle)$.

DEFINITION 4.3 (Riemannian embedding).– *A **Riemannian embedding** is a smooth map $f: M \to N$ between two manifolds $(M, g_{\mu\nu})$ and (N, h_{ij}) such as $g_{\mu\nu} = f^* h_{ij}$, that is, $g(\mathbf{x}, \mathbf{y}) = h(Df(\mathbf{x}), Df(\mathbf{y}))$ for all tangent vectors $\mathbf{x}, \mathbf{y} \in T_pM$ and all $\mathbf{p} \in M$.*

The Nash embedding theorem (Nash 1956) states that any Riemannian metric on any manifold can be realized as the induced metric of some isometric embedding in Euclidean space, thus preserving the length of every path.

DEFINITION 4.4 (Curve length).– *Considering a manifold $(M, g_{\mu\nu})$, **the squared arc length** ds^2 is given by the metric $g_{\mu\nu}$ on this manifold*

$$ds^2 = g_{\mu\nu}dx^\mu dx^\nu \qquad\qquad [4.2]$$

Note the use of Einstein's summation convention, that is, identical indices that appear one up and one down are summed over. For example, if $\mu, \nu = \{1, 2\}$, $i, j = \{1, 2, 3\}$ and $h_{ij} = \delta_{ij}$, then $ds^2 = g_{11}dx^2 + g_{12}dxdy + g_{21}dydx + g_{22}dy^2$.

In the particular case of Euclidean space $\mathbb{R}^3$ with the Cartesian coordinate system (x_0, x_1, x_2), the length element is defined as

$$ds^2 = \delta_{\mu\nu}dx^\mu dx^\nu = dx_0^2 + dx_1^2 + dx_2^2$$

DEFINITION 4.5 (Riemannian gradient).– *Let $f: \mathbb{R}^n \to \mathbb{R}$ be a function such that $f \in C^\infty(M)$ and the differential application $d_x f: T_x M \to \mathbb{R}$ is linear for any $\mathbf{x} \in M$. Thus, there exists a vector field on TM called Riemannian gradient of f and denoted $\nabla_g f$ such that*

$$\langle \nabla_g f, \mathbf{x} \rangle_{g(x)} = d_x f(\mathbf{x}) \quad \text{for all } \mathbf{x} \in T_x M$$

In a local coordinate system $(x_1, \ldots, x_n)$, the expression of the Riemannian gradient is given as

$$\nabla_g f = g^{ij} \frac{\partial f}{\partial x_i} \frac{\partial f}{\partial x_j}$$

where g^{ij} is the inverse of the metric g_{ij}. By noting $G^{-1} = g^{ij}$, the following expression is generally adopted:

$$\nabla_g f = G^{-1} \nabla f$$

When $g_{ij} = \delta_{ij}$ is the identity matrix, the Riemannian gradient $\nabla_g f$ defines the Euclidean gradient ∇f.

DEFINITION 4.6 (Riemannian divergence).– *For each function f with a compact support and a vector field $\mathbf{x} = x^i \partial_i$, the divergence is given by the expression*

$$div\ \mathbf{x} = \frac{1}{\sqrt{\det g}} \partial_i (x^i \sqrt{\det g})$$

where $\det g$ is the determinant of the metric g_{ij}.

DEFINITION 4.7 (Laplace–Beltrami).– *The Laplace–Beltrami operator is defined by $\Delta_g = div \circ \nabla_g$, which is expressed by*

$$\Delta_g = \frac{1}{\sqrt{\det g}} \partial_j (g^{ij} \sqrt{\det g} \partial_i f)$$

Considering $e = \langle \cdot, \cdot \rangle$ the Euclidean metric on $\mathbb{R}^n$, which is defined in this case by the identity matrix $g_{ij} = \delta_{ij}$ for all $\mathbf{x} \in \mathbb{R}^n$ and $i, j = 1 \ldots n$, the Laplace–Beltrami operator becomes the Laplacian on $\mathbb{R}^n$, that is,

$$\Delta_e = \sum_{i=1}^{n} \frac{\partial^2}{\partial x_i^2}.$$

4.3. Adapted image processing by differential calculus on quadratic surfaces

The issue of mirror distortions was investigated by Bogdanova et al. (2005, 2007), who pointed out the need to consider the geometry of the mirror when processing catadioptric images. They propose to include the deformations of the mirror by calculating an adapted Riemannian metric. The main benefit of this approach is to make processing possible directly in the image plane. Following the geometry of the mirror and the image formation path, the authors propose formulating the metric by considering the projection on a Riemannian manifold, namely the surface of the mirror.

4.3.1. *Riemannian geometry for hyperbolic mirrors*

The metric is defined by a stereographic projection from the hyperboloid surface onto the image plane (Figure 4.2). It maps the upper sheet of the hyperboloid onto the open disk. In terms of the (x, y) disk coordinates, the metric on the upper sheet of the unit hyperboloid takes the form

$$(g_{ij})_h = \begin{pmatrix} \dfrac{4}{(1 - x^2 - y^2)^2} & 0 \\ 0 & \dfrac{4}{(1 - x^2 - y^2)^2} \end{pmatrix} \qquad [4.3]$$

The open disk with this metric D^+ is called the Poincaré model of Lobachevsky geometry. The gradient and the Laplacian operators can then be deduced by scaling the corresponding planar Euclidean operators by

$$\nabla_{D+}f = \frac{\left(1 - x^2 - y^2\right)^2}{4}\nabla_{\mathbb{R}^2}f \text{ and } \Delta_{D+}f = \frac{\left(1 - x^2 - y^2\right)^2}{4}\Delta_{\mathbb{R}^2}f \quad [4.4]$$

4.3.2. *Riemannian geometry for spherical mirrors*

The metric is defined by a stereographic projection from the sphere onto the image plane (Figure 4.3). In terms of Euclidean coordinates of the plane (x, y), the metric takes the form

$$(g_{ij})_s = \begin{pmatrix} \dfrac{4}{(1 + x^2 + y^2)^2} & 0 \\ 0 & \dfrac{4}{(1 + x^2 + y^2)^2} \end{pmatrix} \qquad [4.5]$$

Similarly to the previous case, the gradient and Laplacian operators of a spherical catadioptric image are given by

$$\nabla_{\mathbb{S}^2}f = \frac{\left(1 + x^2 + y^2\right)^2}{4}\nabla_{\mathbb{R}^2}f \quad \text{and} \quad \Delta_{\mathbb{S}^2}f = \frac{\left(1 + x^2 + y^2\right)^2}{4}\Delta_{\mathbb{R}^2}f \quad [4.6]$$

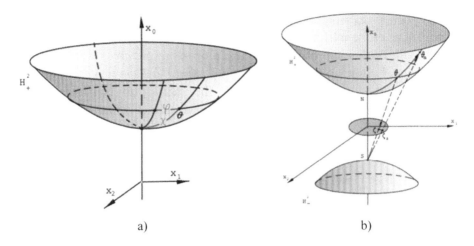

Figure 4.2. *Geometry and image formation on a two-sheet hyperboloid surface: (a) polar coordinates (φ, θ) on the hyperboloid; (b) stereographic projection. For a color version of this figure, see www.iste.co.uk/vasseur/omnidirectional.zip*

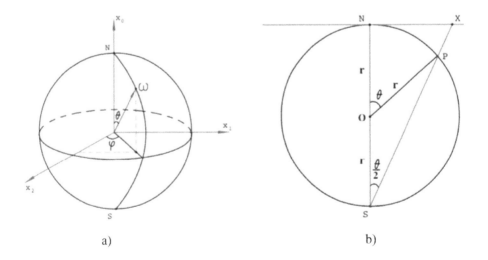

Figure 4.3. *Geometry and image formation on a spherical surface: (a) Polar coordinates (φ, θ) on the sphere; (b) stereographic projection. For a color version of this figure, see www.iste.co.uk/vasseur/omnidirectional.zip*

4.3.3. *Riemannian geometry for paraboloid mirrors*

The metric is defined by an orthographic projection of the paraboloid surface onto the image plane (Figure 4.4). In terms of Euclidean coordinates of the plane (x, y), the metric takes the form

$$(g_{ij})_p = \begin{pmatrix} 1 + 4x^2 & 4xy \\ 4xy & 1 + 4y^2 \end{pmatrix}$$ [4.7]

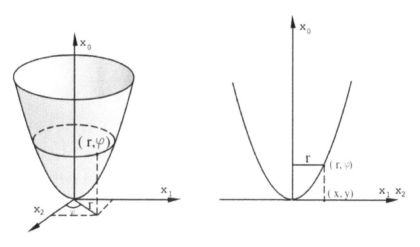

Figure 4.4. *Geometry and image formation on a paraboloid surface. Left: (r, φ)-polar coordinates of the paraboloid mapped to coordinates (x, y) after an orthographic projection, as shown in the image on the right. For a color version of this figure, see www.iste.co.uk/vasseur/omnidirectional.zip*

The calculation for the Laplacian on the paraboloid surface leads to the expression

$$\Delta_{\mathbb{P}^2} f = -\frac{1}{(1 + 4x^2 + 4y^2)} \left[(1 + 4y^2)\frac{\partial^2 f}{\partial x^2} - 4xy\frac{\partial^2 f}{\partial x \partial y} + (1 + 4x^2)\frac{\partial^2 f}{\partial y^2} \right.$$
$$\left. -8x\frac{(1 + 2x^2 + 2y^2)}{1 + 4x^2 + 4y^2}\frac{\partial f}{\partial x} - 8y\frac{(1 + 2x^2 + 2y^2)}{1 + 4x^2 + 4y^2}\frac{\partial f}{\partial y} \right]$$
[4.8]

4.3.4. *Application to active contour deformation*

Parametric active contours describe the evolution of a parametric curve $C(p) = (x(p), y(p))$ with $p \in [0.1]$ in the image domain by minimizing the sum of internal and external energies that control the deformations of the curve

$$E(C) = \int_0^1 E_{int}(C(p)) + E_{ext}(C(p))dp \qquad [4.9]$$

Caselles et al. (1997a, 1997b) showed that the energy minimization problem $E(C)$ [4.9] is in fact a geodesic in a Riemannian space with the metric: $g_{ij} = f^2(|\nabla I|)\delta_{ij}$, such that,

$$E(C) = \int_0^1 \sqrt{g_{ij}C_i'C_j'}dp = \int_0^1 \sqrt{f^2(|\nabla I|)\delta_{ij}C_i'C_j'}dp \qquad [4.10]$$

$$= \int_0^1 f(|\nabla I(C(p))|)|C'(p)|dp \qquad [4.11]$$

The problem can then be interpreted as the minimization of the length of the contour or the surface in a space endowed with a metric taking into account the characteristics of the image. The classical active contour model uses in its formulation the Euclidean metric δ_{ij} [4.10]. For catadioptric omnidirectional images, (Bogdanova et al. 2007) have developed a model for spherical, paraboloid and hyperboloid images using Riemannian metrics g_{ij} conformal and equivalent to the Euclidean metric, computed using the stereographic projection. Thus, the energy of the standard model [4.10] becomes in the catadioptric case

$$E(C) = \int_0^1 \sqrt{f^2(|\nabla I|)g_{ij}C_i'C_j'}dp \qquad [4.12]$$

$$= \int_0^1 f(|\nabla I(C(p))|)|g|^{1/4}|C'(p)|dp \qquad [4.13]$$

where $|g|$ is the determinant of the Riemannian metric g_{ij}. For spherical and paraboloid images, this metric is given by

$$g_{ij} = \begin{bmatrix} \frac{4}{(1+x^2+y^2)^2} & 0 \\ 0 & \frac{4}{(1+x^2+y^2)^2} \end{bmatrix} \qquad [4.14]$$

and, in the case of hyperboloid images, it is written as

$$g_{ij} = \begin{bmatrix} \frac{4}{(1-(x^2+y^2))^2} & 0 \\ 0 & \frac{4}{(1-(x^2+y^2))^2} \end{bmatrix} \qquad [4.15]$$

Applied to object segmentation, the effect of the hyperbolic metric in comparison to the Euclidean one is shown in Figure 4.5. We can notice a smooth transition on the hyperbolic manifold, while it appears as a fast transition on the hyperbolic plane image. The results show that the method based on the Riemannian metric succeeds in segmenting the object of interest while the Euclidean metric fails to segment the entire object of interest. This is due to the false contour corresponding to the fast transition on the hyperboloid image. The weighting factor of the Riemannian metric leads to smooth this transition in $f(|\nabla I(C(p))|)$ to end on a good indicator of the real contour of the object.

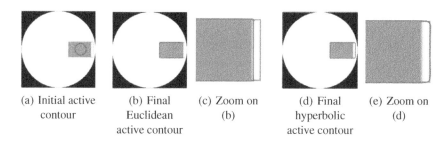

| (a) Initial active contour | (b) Final Euclidean active contour | (c) Zoom on (b) | (d) Final hyperbolic active contour | (e) Zoom on (d) |

Figure 4.5. *Segmentation of an object of interest in a synthetic hyperboloid image using classic Euclidean active contour method (b) and (c), and using the induced Riemannian metric from the mirror geometry in (d) and (e) (source: Bogdanova et al. 2007). For a color version of this figure, see www.iste.co.uk/vasseur/omnidirectional.zip*

4.4. Adapted image processing by Riemannian geodesic metrics

In recent literature, Aziz et al. (2018) have proposed a generic formulation of Riemannian geodesic metrics for central catadioptric color images, derived using the unified equivalence sphere model. It is therefore straightforward to define the metric of any central catadioptric system (parabolic, hyperbolic, etc.) by simply selecting its associated mirror parameters. This formulation allows direct processing on the plane of the original catadioptric image without the need to wrap it onto the surface of the equivalence sphere, thus preserving the whole visual information during processing. The embedded metric is called a hybrid geodesic metric, as it combines both the spatial and the color components of the catadioptric image.

Embedding an image in a higher dimensional space has been studied in the literature for conventional images (Sochen and Zeevi 1998; Kimmel et al. 1998), where a greyscale image is considered as a surface (x, y, I) in 3D space as illustrated in Figure 4.6, and a color image (x, y, R, G, B) in 5D space. This relevant embedding allows the use of more general operators like Laplace–Beltrami on manifolds (Wetzler et al. 2013). Furthermore, it yields to efficient methods for image enhancement and denoising (Spira et al. 2005, 2007).

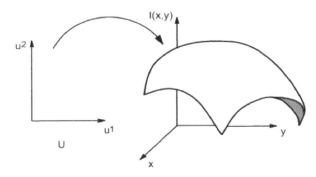

Figure 4.6. *Embedding a 2D image as a surface in a three-dimensional space*

For color catadioptric images, Aziz et al. (2018) consider a Riemannian embedding, mapping the image plane $\mathbb{R}^2$ to a $(m + 3)$ dimensional manifold

$$f: \qquad \mathbb{R}^2 \to \mathbb{R}^{3+m} \qquad\qquad\qquad [4.16]$$

$$(x, y) \mapsto \left(\chi_1^{-1}, \chi_2^{-1}, \chi_3^{-1}, \beta I^1, \dots, \beta I^m \right)$$

The embedding space is composed of two parts, the first represents the inverse mapping on the sphere $\chi^{-1}(x, y)$ and the second part is related to the m-components of the image intensity. Since spatial coordinates and image photometric components in [4.16] do not induce the same measure, the parameter β acts as a relative scaling factor in the space-feature manifold $(\mathbb{R}^{3+m})$.

By construction, the Riemannian embedding [4.16] assumes that the spatial-color metric space $\mathbb{R}^{m+3}$ is endowed with a Euclidean metric, denoted h_{ij}. The catadioptric image plane $\mathbb{R}^2$ is non-Euclidean: its Riemannian metric, denoted $g_{\mu\nu}$, is induced by h_{ij} and called *Pullback metric*. Formally, this *Pullback metric* is a symmetric two-tensor defined as

$$g_{\mu\nu} = h_{ij}\partial_\mu f^i \partial_\nu f^j \qquad\qquad [4.17]$$

where $\mu, \nu = \{1, 2\}$ and $i, j = \{1, \ldots, 3 + m\}$. $\partial_\mu f^i$ is the partial derivative of f^i with respect to μ variable.

4.4.1. *Spatial Riemannian metric*

Considering only the spatial geometric embedding, we can compute the spatial geometric metric on the catadioptric image plane using the parameters of the equivalence sphere model

$$g_{\mu\nu} = \zeta \begin{bmatrix} 1 + \left(1 - \xi^2\right) y^2 & -\left(1 - \xi^2\right) xy \\ -\left(1 - \xi^2\right) xy & 1 + \left(1 - \xi^2\right) x^2 \end{bmatrix} \qquad\qquad [4.18]$$

with

$$\zeta = \frac{\left(\xi + \sqrt{1 + (1 - \xi^2)(x^2 + y^2)}\right)^2}{(x^2 + y^2 + 1)^2 \left(1 + (1 - \xi^2)(x^2 + y^2)\right)}. \qquad\qquad [4.19]$$

This is a generic formulation, which can be made explicit according to the catadioptric mirror models.

REMARK 4.1.– Para-catadioptric projection $\xi = 1$

This projection corresponds to a parabolic mirror and is stereographic (i.e. of the north pole of the sphere). It provides a conformal metric[1] to the Euclidean metric for the catadioptric image plane

$$(g_{\mu\nu})_{para} = \begin{bmatrix} \frac{4}{(x^2+y^2+1)^2} & 0 \\ 0 & \frac{4}{(x^2+y^2+1)^2} \end{bmatrix}. \qquad\qquad [4.20]$$

---

1 The conformal property means that this projection preserves the angles between intersecting curves.

Note that the metric in this case [4.20] yields to the same Riemannian metric expression introduced in Bogdanova et al. (2007) and Aziz et al. (2016).

REMARK 4.2.– Hyper-catadioptric projection $0 < \xi < 1$

This is a non-conformal projection and corresponds to the case of hyperbolic and elliptical mirrors. It depends on the ξ parameter. For example, taking $\xi = 0.7$ leads to the Riemannian metric tensor

$$(g_{\mu\nu})_{hyper} =$$

$$\frac{\left(0.7 + \sqrt{1 + 0.51(x^2 + y^2)}\right)^2}{(x^2 + y^2 + 1)^2 \left(1 + 0.51(x^2 + y^2)\right)} \begin{bmatrix} 1 + 0.51y^2 & -0.51xy \\ -0.51xy & 1 + 0.51x^2 \end{bmatrix} . \quad [4.21]$$

REMARK 4.3.– Perspective projection $\xi = 0$

This provides a non-conformal projection that corresponds to the case of planar mirrors

$$(g_{\mu\nu})_{pers} = \frac{1}{(x^2 + y^2 + 1)^2} \begin{bmatrix} 1 + y^2 & -xy \\ -xy & 1 + x^2 \end{bmatrix} . \quad [4.22]$$

4.4.2. *Spatial-color metric*

Besides the spatial embedding, the generic hybrid embedding also integrates the colorimetric components of the image. This embedding is expressed in a generic form [4.16]

$$g_{\mu\nu} =$$

$$\begin{bmatrix} \zeta\left(1 + (1 - \xi^2)y^2\right) + \beta^2 \sum_{i=1}^{m}(I_x^i)^2 & -\zeta(1 - \xi^2)xy + \beta^2 \sum_{i=1}^{m} I_x^i I_y^i \\ -\zeta(1 - \xi^2)xy + \beta^2 \sum_{i=1}^{m} I_x^i I_y^i & \zeta\left(1 + (1 - \xi^2)x^2\right) + \beta^2 \sum_{i=1}^{m}(I_y^i)^2 \end{bmatrix}$$

$$[4.23]$$

with ζ as given above by [4.19].

It should be noted that the catadioptric spatial distortion and/or the multi-component photometric information of the image can be considered simultaneously or separately. For instance, considering only the spatial aspect, it is equivalent to metric [4.18]. In the particular case where we consider only the colorimetric space, we get

$$f : \quad \mathbb{R}^2 \to \mathbb{R}^m$$
$$(x, y) \mapsto (I^1, \dots, I^m)$$

[4.24]

which corresponds to the color tensor in Di Zenzo's approach (Di Zenzo 1986)

$$g_{\mu\nu} = \begin{bmatrix} \sum_{i=1}^{m} (I_x^i)^2 & \sum_{i=1}^{m} I_x^i I_y^i \\ \sum_{i=1}^{m} I_x^i I_y^i & \sum_{i=1}^{m} (I_y^i)^2 \end{bmatrix}.$$

[4.25]

4.4.3. *Application to Gaussian kernel based smoothing*

Gaussian kernels are widely used in image processing. Aziz et al. (2018) have proposed a Riemannian formulation of the Gaussian kernel for catadioptric images, built as a solution for the diffusion equation in terms of the Laplace–Beltrami operator using the previously defined metrics.

$$G_g(\mathbf{x}, \sigma) = \frac{1}{2\pi\sigma^2 |g|^{-1/2}} \exp\left(-\frac{1}{2\sigma^2} \mathbf{x}^T g_{\mu\nu} \mathbf{x} \right)$$

[4.26]

where $|g| = det(g_{\mu\nu})$ is the determinant of the tensor metric $g_{\mu\nu}$, and $\mathbf{x} = (x, y)^T$ is the point position in the catadioptric plane.

The effect of the metric in the smoothing process is crucial as shown in Figure 4.7 using synthetic images obtained by a para-catadioptric sensor ($\xi = 1$). With the Euclidean metric, the same blur is produced uniformly over the entire image. With the induced spatial Riemannian metric, the weight of the Gaussian function depends on the position of each image point. Consequently, a suitable smoothing is obtained which is more significantly perceptible near the image center and the edges are more preserved. Indeed, the color part of the hybrid metric acts as an edge indicator having a good edge-preserving property, while the anisotropic Gaussian function handles the nonuniform smoothing from the periphery to the center because of the spatial part of the metric.

Formulation [4.26] can then be easily extended to express the Difference of Gaussians (DoG) operator for non-Euclidean images. The latter is used in practice in many applications in scale-space theory, interest feature extraction and edge detection and sharpening.

$$DoG_g(\mathbf{x}, \sigma_1, \sigma_2) = G_g(\mathbf{x}, \sigma_1) - G_g(\mathbf{x}, \sigma_2) \hspace{2cm} [4.27]$$

$$= \frac{1}{2\pi|g|^{-1/2}} \left[\frac{1}{\sigma_1^2} \exp\left(-\frac{1}{2\sigma_1^2} \mathbf{x}^T g_{\mu\nu} \mathbf{x} \right) - \frac{1}{\sigma_2^2} \exp\left(-\frac{1}{2\sigma_2^2} \mathbf{x}^T g_{\mu\nu} \mathbf{x} \right) \right]$$

where $\sigma_1 > \sigma_2$ and commonly $\sigma_1 = \sqrt{2}\sigma_2$. Figure 4.8 demonstrates the potential of the adapted DoG operator, using synthetic para-catadioptric ($\xi = 1$) and real hyper-catadioptric ($\xi = 0.91$) images. These images are convolved, on the one hand, with the classical DoG, using standard Euclidean Gaussian kernel, and, on the other hand, with the adapted DoG_g based on the Riemannian-adapted Gaussian kernel. We can clearly see the improvement produced by the adapted DoG_g operator, which yields extracting edges with more accurate details and a more regular thickness.

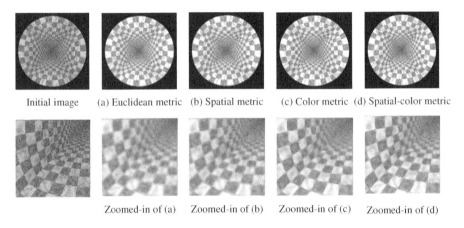

Initial image   (a) Euclidean metric   (b) Spatial metric   (c) Color metric   (d) Spatial-color metric

Zoomed-in of (a)   Zoomed-in of (b)   Zoomed-in of (c)   Zoomed-in of (d)

Figure 4.7. *Adapted Gaussian smoothing. On the top row: smoothing using different metrics (Euclidean, spatial, color and spatial-color metrics respectively). On the bottom row: the corresponding zoomed-in views. For a color version of this figure, see www.iste.co.uk/vasseur/omnidirectional.zip*

4.4.4. *Application to corner features detection*

Based on the hybrid Riemannian tensor and the adapted Gaussian kernel, Aziz et al. (2018) have also proposed an adaptive Harris corner detector for catadioptric images. The adaptation concerns the calculation of the components of the structure tensor, where, in particular, the spatial derivatives of the image are computed using the adapted Gaussian kernel derivatives, and the adaptive Gaussian smoothing of the structure tensor elements is used in order to obtain stable and robust interest points. The effect of the metric in corner detection is depicted in Figure 4.9 using examples of synthetic, real para- and hyper-catadioptric images.

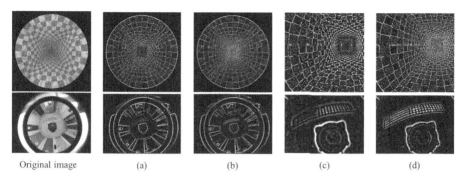

Original image   (a)   (b)   (c)   (d)

Figure 4.8. *Adapted Difference of Gaussians filtering: (a) Classical approach: Euclidean DoG. (b) Riemannian approach based on the Riemannian spatial metric. (c) and (d) the zoomed-in views near the central regions of (a) and (b) respectively. For a color version of this figure, see www.iste.co.uk/vasseur/omnidirectional.zip*

4.5. Adapted image processing by spherical geodesic distance

In perspective image processing, most methods are developed from Euclidean metric

$$\forall x, y \in \mathbb{R}^2, d_{\mathbb{R}^2}(x, y) = ||x - y||_{l^2(\mathbb{R}^2)} \qquad [4.28]$$

From this metric, a neighborhood can be derived to define pixels dependency between each other. Derivative computation, corner detection, or point matching by correlation can then be performed because of this neighborhood definition. Theoretically, the neighborhood expresses the mutual 3D point influence into the image plane. Practically, if we consider an orthographic camera, the Euclidean neighborhood represents exactly 3D points placed on a fronto-parallel plane (Figure 4.10(a)). In the case of a perspective camera, this neighborhood is generally considered as a sufficiently good approximation whatever the 3D point configuration (Figure 4.10(b)). Considering a central catadioptric image, the Euclidean neighborhood is no longer relevant (Figure 4.10(c)). However, a regular neighborhood according to θ and ϕ angles (Figure 4.10(d)) will translate exactly mutual 3D point influence if they are placed on a concentric sphere of the unit sphere. This last neighborhood is developed in the following.

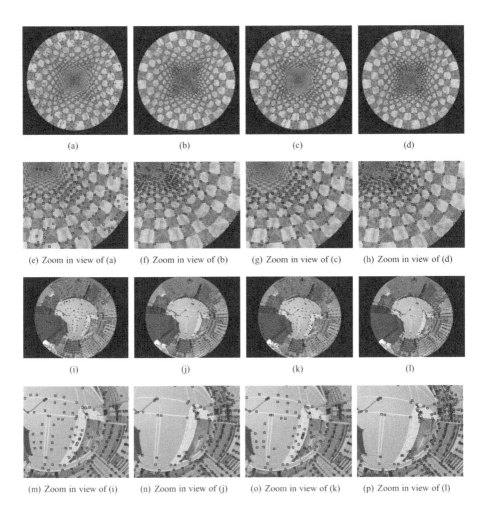

Figure 4.9. *Harris corner detection using the adapted Gaussian kernel with different metrics (600 strongest corners): the Euclidean metric that leads to the classical Euclidean corner detection in (a) and (i); the Riemannian spatial metric that takes into account the mirror deformation in (b) and (j). Using the color metric only in (c) and (k), and the hybrid spatial-color metric that combines both the color information and the mirror distortion in (d) and (l). For a color version of this figure, see www.iste.co.uk/vasseur/omnidirectional.zip*

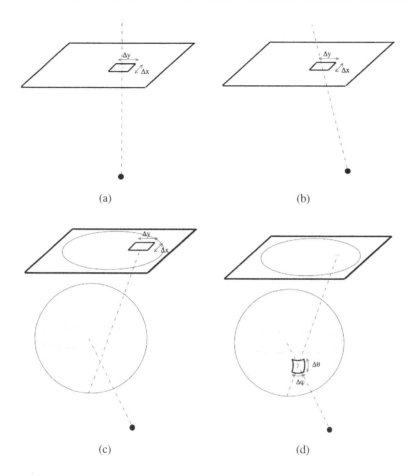

Figure 4.10. *Image formation and neighborhood dependency. (a) Orthographic image, Euclidean metric; (b) perspective image, Euclidean metric; (c) omnidirectional image, Euclidean metric; (d) omnidirectional image, geodesic metric*

4.5.1. *Neighborhood definition*

Let P be the projection that transforms a central catadioptric image which belongs to $\mathbb{R}^2$ into its equivalent spherical image on $\mathcal{S}^2$

$$
\mathrm{P}: \begin{array}{ccc} \mathbb{R}^2 & \to & \mathcal{S}^2 \\ \mathbf{x} & \mapsto & \mathbf{x}_s = (\theta, \phi) \end{array}
\qquad [4.29]
$$

In order to process spherical images, Demonceaux et al. (2011) and Demonceaux and Vasseur (2006) have proposed the use of the following geodesic distance:

$$\forall \mathbf{x}_s = (\theta, \phi), \mathbf{y}_s = (\theta', \phi') \in \mathcal{S}^2$$

$$d_{\mathcal{S}^2}(\mathbf{x}, \mathbf{y}) = \arccos \left[\begin{pmatrix} \cos(\phi)\sin(\theta) \\ \sin(\phi)\sin(\theta) \\ \cos(\theta) \end{pmatrix}^T \cdot \begin{pmatrix} \cos(\phi')\sin(\theta') \\ \sin(\phi')\sin(\theta') \\ \cos(\theta') \end{pmatrix} \right]$$

Let $\mathbf{x}_s \in \mathcal{S}^2$ be the projection of an image pixel $\mathbf{x}$ onto the unitary sphere ($\mathbf{x}_s = \mathrm{P}(\mathbf{x})$). The continuous neighborhood $\mathcal{V}_r(\mathbf{x})$ of pixel $\mathbf{x}$ is defined in the image as follows:

$$\mathcal{V}_r(\mathbf{x}) = \left\{ \mathbf{y}_s \in \mathcal{S}^2, d_{\mathcal{S}^2}(\mathbf{x}_s, \mathbf{y}_s) \leqslant r \,|\, \mathrm{P}(\mathbf{x}) = \mathbf{x}_s \right\} \qquad [4.30]$$

This neighborhood results from the intersection between the cone with apex O, direction $O\mathbf{x}_s$, angle r and the sphere $\mathcal{S}^2$. To use discrete convolution filters with this neighborhood, it is then necessary to introduce a discrete version of [4.30]. In this way, the tangent plane π to $\mathcal{S}^2$ at $\mathbf{x}_s$ is considered. This plane has the following equation: $\mathbf{x}_s + a\vec{e}_\theta + b\vec{e}_\phi$, $(a, b) \in \mathbb{R}^2$, where $(\vec{e}_\theta, \vec{e}_\phi)$ are vectors of the basis at $\mathbf{x}_s = (\theta, \phi)$ (Figure 4.11)

$$\vec{e}_\theta = \frac{\partial O\vec{M}}{\partial \theta} = \begin{pmatrix} \cos\phi\cos\theta \\ \sin\phi\cos\theta \\ -\sin\theta \end{pmatrix}, \quad \vec{e}_\phi = \frac{1}{\sin\theta}\frac{\partial O\vec{M}}{\partial \phi} = \begin{pmatrix} -\sin\phi \\ \cos\phi \\ 0 \end{pmatrix}$$

To have a regular mesh with constant geodesic distance, $(2N+1)^2$ points of the tangent plane π are obtained as follows:

$$\mathbf{x}_s + \tan(nr)\vec{e}_\theta + \tan(pr)\vec{e}_\phi, \quad -N \leq n, p \leq N \qquad [4.31]$$

i.e. $(2N+1)^2$ points of $\mathcal{S}^2$ as

$$\mathcal{V}_r^N(\mathbf{x}) = \left\{ \mathbf{x}_s(n, p) = \frac{\mathbf{x}_s + \tan(nr)\vec{e}_\theta + \tan(pr)\vec{e}_\phi}{\|\mathbf{x}_s + \tan(nr)\vec{e}_\theta + \tan(pr)\vec{e}_\phi\|}, \right.$$

$$\left. -N \leq n, p \leq N, \mathbf{x}_s = \mathrm{P}(\mathbf{x}) \right\} \qquad [4.32]$$

In $\mathcal{V}_r^N(\mathbf{x})$, the points are all equidistant in terms of the geodesic metric

$$\forall \mathbf{y} \in \mathcal{V}_r^N(\mathbf{x}), \quad \min_{\mathbf{z} \in \mathcal{V}_r^N(\mathbf{x}) \setminus \mathbf{y}} d_{\mathcal{S}^2}(\mathbf{y}, \mathbf{z}) = r$$

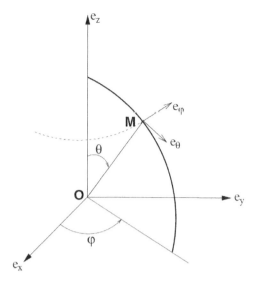

Figure 4.11. *Spherical coordinates and local basis*

4.5.2. *Application to linear catadioptric image filtering*

Linear image filtering consists of applying a weighted mask at each image point through a convolution product. According to the previously mentioned reasons, the conventional convolution product is not applicable to catadioptric images. Consequently, the first step consists of defining the convolution product based on the geodesic metric in order to process each pixel of the catadioptric image. It has been shown (Demonceaux et al. 2011) that this new definition implies that the conventional operators remain valid.

Consider an omnidirectional image I and denote $I_{\mathcal{V}_r^N(\mathbf{x})} = I(\mathrm{P}^{-1}(\mathcal{V}_r^N(\mathbf{x})))$, where $I_{\mathcal{V}_r^N(\mathbf{x})}$ represents the $(2N+1)^2$ grey level values of the regular grid centered at $\mathbf{x}_s = \mathrm{P}(\mathbf{x})$ (where P is defined by equation [4.29]).

$$I_{\mathcal{V}_r^N(\mathbf{x})}(n,p) = I(\mathrm{P}^{-1}(\mathbf{x}_s(n,p)))\quad -N \leq n, p \leq N$$

Let H be a filter with size $(2N + 1) \times (2N + 1)$, the convolution product of the image I by the filter H at a point $\mathbf{x} \in \mathbb{R}^2$ is defined as follows:

$$\forall \mathbf{x} \in \mathbb{R}^2, IH(\mathbf{x}) = I_{\mathcal{V}_r^N(\mathbf{x})} * H(\mathbf{x})$$
$$= \sum_{i=-N}^{N} \sum_{j=-N}^{N} I_{\mathcal{V}_r^N(\mathbf{x})}(i, j) H(i, j) \qquad [4.33]$$

This new definition of convolution takes into account the distortions and offers a new paradigm for omnidirectional image processing. Indeed, this formulation can easily reformulate classical filters as derivative filters, feature detection and matching, etc.

For instance, a classical Sobel filter can be described as

$$\|\nabla I(\mathbf{x}_s)\|^2 \simeq |I_{\mathcal{V}_r^1(\mathbf{x})} * S|^2 + |I_{\mathcal{V}_r^1(\mathbf{x})} * S^T|^2$$

where

$$S = \begin{pmatrix} -1 & -2 & -1 \\ 0 & 0 & 0 \\ 1 & 2 & 1 \end{pmatrix}$$

In order to verify the behavior of this geodesic filtering, we can apply it to a synthetic image and compare the results with the classical Sobel filter (Figure 4.12) and with Bogdanova's method (Bogdanova et al. 2007). This synthetic image simulates a catadioptric camera placed inside a rectangular parallelepiped. This configuration is not optimal for the geodesic neighborhood. Indeed, as mentioned before, the geodesic neighborhood is optimal for the case of a spherical scene captured from its center by a catadioptric camera. Despite this, a slight qualitative improvement of the gradient computation can be noted. The edges are more clearly distinguished in the center of the image and their thickness is more regular along the radial lines in the geodesic approach (Figure 4.12(f)). Note that Bogdanova's approach globally improves detection but reduces the detection of lines near the image center (Figure 4.12(c)).

Similarly to the previous case, Gaussian and Laplacian filters remain valid with the geodesic neighborhood. Indeed, we can consider the Gaussian filtering as a weighting of the points according to their distance with the central point. This leads to substitute the Gaussian

$$G_{\mathbf{x}}(\mathbf{y}, \sigma) = \frac{1}{2\pi\sigma^2} e^{-\frac{||\mathbf{x} - \mathbf{y}||^2}{2\sigma^2}} \quad (\mathbf{x}, \mathbf{y}) \in \mathbb{R}^2$$

by

$$G_{\mathbf{x}}(\mathbf{y}, \sigma) = \frac{1}{2\pi\sigma^2} e^{-\frac{d_{\mathcal{S}^2}(\mathbf{x}, \mathbf{y})}{2\sigma^2}} \quad (\mathbf{x}, \mathbf{y}) \in \mathcal{S}^2. \qquad [4.34]$$

The Laplacian of Gaussian can also be defined as

$$LoG_{\mathbf{x}}(\mathbf{y}) = -\frac{1}{\pi\sigma^4} \left[1 - \frac{d_{\mathcal{S}^2}(\mathbf{x}, \mathbf{y})}{2\sigma^2} \right] \exp -\frac{d_{\mathcal{S}^2}(\mathbf{x}, \mathbf{y})}{2\sigma^2} \qquad [4.35]$$

Thus, while the Gaussian or Laplacian of Gaussian definitions using spherical harmonic analysis for spherical image processing require their own reformulation (Bülow 2002; Daniilidis et al. 2002), the geodesic approach provides a straightforward adaptation to such conventional filters. Indeed, because of equations [4.34], [4.35] and the convolution product defined in equation [4.33], the kernels used in perspective image processing remain valid.

4.5.3. *Application to corner features detection and matching*

The behavior of these filters can also be tested on the Harris corner detector. In perspective image processing, the corner detection by the Harris method consists of studying the eigenvalues of matrix M at each image point $\mathbf{x}$

$$M(\mathbf{x}) = \begin{bmatrix} L_x^2(\mathbf{x}, \sigma) & L_x L_y(\mathbf{x}, \sigma) \\ L_x L_y(\mathbf{x}, \sigma) & L_y^2(\mathbf{x}, \sigma) \end{bmatrix} \qquad [4.36]$$

where $L_i(\mathbf{x}, \sigma) = \frac{\partial}{\partial i} G_{\mathbf{x}}(\mathbf{x}, \sigma) * I(\mathbf{x})$.

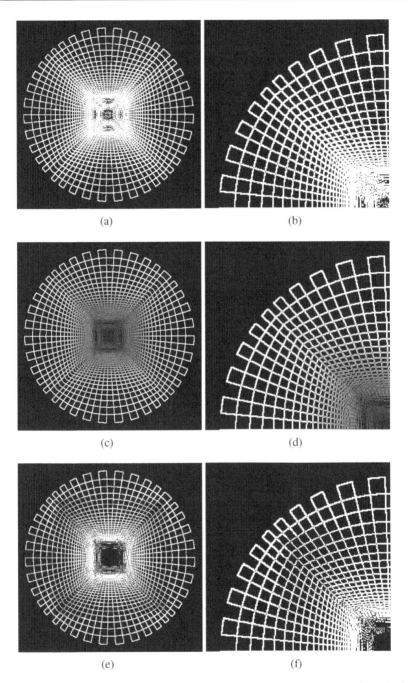

(a)

(b)

(c)

(d)

(e)

(f)

Figure 4.12. *Gradient norm results on a synthetic image. (a and b) Classical perspective processing approach; (c and d) Bogdanova's method (Bogdanova et al. 2007); (e and f) spherical geodesic approach*

This matrix M can be computed with classical filter, with spherical convolution and spherical Gaussian function (Bülow 2002). The adapted Harris corner detection with a classical Harris and a Harris computed on the sphere, for the synthetic image, are shown in Figure 4.13. Figure 4.13(b) corresponds to the detection using classical Harris; some corner detections are double while some other corners are not detected (17 corners have been detected). Conversely, Harris detection by spherical analysis detects 21 corners with more corners at the image center where there is a blind spot in the real image (Figure 4.13(a)), but does not detect corners at the periphery (Figure 4.13(c)). Thus, the adapted convolution seems to be a good compromise (Figure 4.13(d)): 19 corners have been detected and the results show that the geodesic distance seems capable to correctly detect corners with only slight modifications of the conventional technique.

This formulation can extend to feature points matching. Indeed, similarly to image filtering methods, matching techniques must take into account the deformations of the image if they are based on the neighborhood. Consider the zero mean normalized cross-correlation (ZNCC) function

$$ZNCC(\mathbf{x}, \mathbf{y}) = \frac{\sum_{\mathbf{i} \in \mathcal{V}(\mathbf{x})} \sum_{\mathbf{j} \in \mathcal{V}(\mathbf{y})} (I_1(\mathbf{i}) - \overline{I_1(\mathbf{x})})(I_2(\mathbf{j}) - \overline{I_2(\mathbf{y})})}{\sqrt{\sum_{\mathbf{i} \in \mathcal{V}(\mathbf{x})} (I_1(\mathbf{i}) - \overline{I_1(\mathbf{x})})^2 \sum_{\mathbf{j} \in \mathcal{V}(\mathbf{y})} (I_2(\mathbf{j}) - \overline{I_2(\mathbf{y})})^2}} \quad [4.37]$$

where $\overline{I_1(\mathbf{x})}$ and $\overline{I_2(\mathbf{y})}$ are the grey level mean of image I_1 and I_2 in neighborhood $\mathcal{V}(\mathbf{x})$ and $\mathcal{V}(\mathbf{y})$), respectively.

For a perspective image, the neighborhood is once again defined from the Euclidean distance between points [4.28]. However, for the same reasons as previously exposed, this neighborhood is not applicable in catadioptric images. Moreover, a neighborhood on the unit sphere defined from a regular sampling of the spherical coordinates θ and ϕ is also not adapted. Indeed, such a definition does not provide the same neighborhood for each point on the grid (Figure 4.10).

By comparing the previous images using a classical neighborhood of size 7×7 and a geodesic neighborhood $\mathcal{V}_r^3$, and considering the same corners in both cases obtained by the classical Harris detector with the same thresholds, the use of the classic ZNCC allows us to match a total of 65 points with 53 correct matchings, which represents a rate of outliers equal to 18.4%. Comparatively, the geodesic neighborhood [4.32] provides 71 matchings including 63 positive matching, that is, a rate equal to 11.2% of outliers (Figure 4.14).

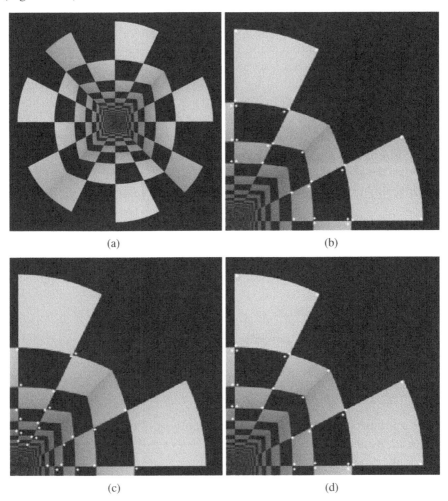

(a)

(b)

(c)

(d)

Figure 4.13. *Harris corner detection. (a) Test Image, (b) classical Harris corner, (c) spherical Harris corner and (d) adapted method. For a color version of this figure, see www.iste.co.uk/vasseur/omnidirectional.zip*

Figure 4.15(c) compares some results for corner detection and matching over an outdoor sequence of 21 consecutive images (Figures 4.15(a) and 4.15(b)). Unlike previous matching results obtained from the classical Harris corner detector, conventional detection and matching are compared to the adapted detection and matching results. In this way, the benefits of the geodesic neighborhood are more significantly exhibited. Green bars (respectively red) correspond to the number of matched corners in each image, while the blue bars (respectively black) describe the number of outliers for each method. It is worth noting that the geodesic approach provides a better matching since the number of matched corners is greater while the number of outliers is less than for the classical method. Indeed, over the 21 images, the geodesic method provides 152 matchings on average while the classical method gives 116 matchings. Also, the outlier rate is equal to 3.5 in the geodesic case, while the classical approach presents a rate of 14.5.

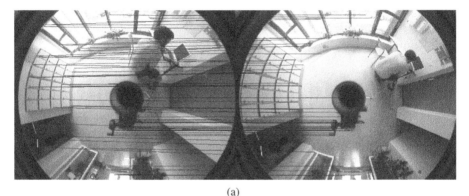

(a)

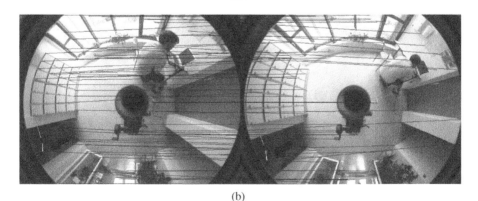

(b)

Figure 4.14. *Matching by ZNCC. (a) Classical method: 65 total matchings and 53 correct matchings. (b) Geodesic method: 71 total matchings and 63 correct matchings. For a color version of this figure, see www.iste.co.uk/vasseur/ omnidirectional.zip*

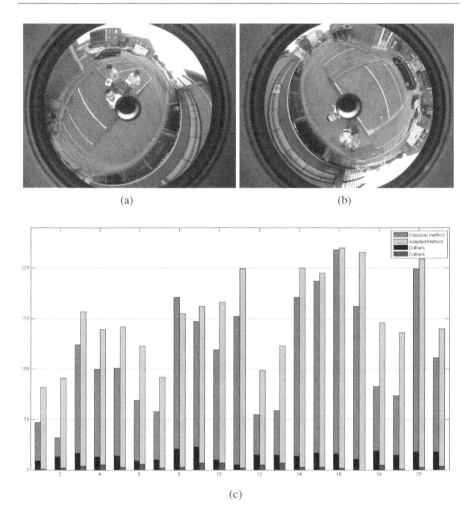

(a)   (b)

(c)

Figure 4.15. *Corner detection and matching. (a and b) Two images from the real sequence. (c) Matching results. For a color version of this figure, see www.iste.co.uk/vasseur/ omnidirectional.zip*

4.6. Conclusion

In this chapter, we have presented different processing approaches adapted to the characteristics and anamorphosis of catadioptric images. The effect of projective mirrors introduces significant biases in the interpretation of visual information and the accurate extraction of features in catadioptric imaging. At the cost of mathematical developments, using sometimes complex theoretical tools, solutions are now possible to go beyond simply applying perspective image processing. These formulations and adaptations allow for more efficient image processing while maintaining the complete information of the image, operating directly on the original catadioptric image plane, and thus avoiding rectifications, interpolations or image transformations.

Cutting-edge work on this topic has shown significant improvements in the illustrated applications of image smoothing, Gaussian difference filtering, edge detection, linear filtering, interest point extraction and scale-space analysis.

4.7. References

Aziz, F., Labbani-Igbida, O., Radgui, A., Tamtaoui, A. (2016). Color-metric tensor for catadioptric systems. In *Proc. IEEE International Conference on Image Processing (ICIP)*, 1594–1598. IEEE Computer Society, Phoenix.

Aziz, F., Labbani-Igbida, O., Radgui, A., Tamtaoui, A. (2018). Generic spatial-color metric for scale-space processing of catadioptric images. *Comput. Vis. Image Underst.*, 176, 54–69.

Baker, S. and Nayar, S. (1999). A theory of single-viewpoint catadioptric image formation. *Int. J. Comput. Vision*, 35(2), 175–196.

Bogdanova, I., Bresson, X., Thiran, J.-P., Vandergheynst, P. (2005). Laplacian operator, diffusion flow and active contour on non-Euclidean images. Technical Report, EPFL, Ecublens.

Bogdanova, I., Bresson, X., Thiran, J.-P., Vandergheynst, P. (2007). Scale space analysis and active contours for omnidirectional images. *IEEE Trans. Image Process.*, 16(7), 1888–1901.

Bülow, T. (2002). Multiscale image processing on the sphere. In *Joint Pattern Recognition Symposium*. Springer, Berlin/Heidelberg.

Caselles, V., Kimmel, R., Sapiro, G. (1997a). Geodesic active contours. *Int. J. Comput. Vision*, 22(1), 61–79.

Caselles, V., Kimmel, R., Sapiro, G., Sbert, C. (1997b). Minimal surfaces based object segmentation. *IEEE Trans. Pattern Anal. Mach. Intell.*, 19(4), 394–398.

Daniilidis, K., Makadia, A., Bulow, T. (2002). Image processing in catadioptric planes: Spatiotemporal derivatives and optical flow computation. In *Proc. of the IEEE Workshop on Omnidirectional Vision*. Held in conjunction with ECCV'02, 3–10. IEEE Computer Society, USA.

Demonceaux, C. and Vasseur, P. (2006). Markov random fields for catadioptric image processing. *Pattern Recognit. Lett.*, 27(16), 1957–1967.

Demonceaux, C., Vasseur, P., Fougerolle, Y. (2011). Central catadioptric image processing with geodesic metric. *Image Vision Comput.*, 29(12), 840–849.

Di Zenzo, S. (1986). A note on the gradient of a multi-image. *Computer Vision, Graphics, and Image Processing*, 33(1), 116–125.

Geyer, C. and Daniilidis, K. (2000). A unifying theory for central panoramic systems and practical implications. In *Proc. 6th Eur. Conf. Comput. Vis.*, 445–461.

Kimmel, R., Malladi, R., Sochen, N. (1997). Image processing via the Beltrami operator. In *Asian Conference on Computer Vision*, 574–581. Springer, Berlin/Heidelberg.

Lee, J.M. (2006). *Riemannian Manifolds: An Introduction to Curvature*. Springer, New York.

Nash, J. (1956). The imbedding problem for Riemannian manifolds. *Annals of Mathematics*, 63(1), 20–63.

Petersen, P. (2006). *Riemannian Geometry*. Springer, New York.

Sochen, N. and Zeevi, Y.Y. (1998). Representation of colored images by manifolds embedded in higher dimensional non-Euclidean space. In *Proc. IEEE International Conference on Image Processing (ICIP)*, 1, 166–170. IEEE Computer Society, Chicago.

Spira, A., Sochen, N., Kimmel, R. (2005). Geometric filters, diffusion flows, and kernels in image processing. In *Handbook of Geometric Computing*, Bayro-Corrochano, E. (ed.). Springer, Berlin/Heidelberg.

Spira, A., Kimmel, R., Sochen, N. (2007). A short-time Beltrami kernel for smoothing images and manifolds. *IEEE Trans. Image Process.*, 16(6), 1628–1636.

Wetzler, A., Aflalo, Y., Dubrovina, A., Kimmel, R. (2013). The Laplace–Beltrami operator: A ubiquitous tool for image and shape processing. In *International Symposium on Mathematical Morphology and Its Applications to Signal and Image Processing*, Springer, Berlin/Heidelberg.

5

Non-Central Sensors and Robot Vision

Sio-hoi IENG

Institut de la Vision, University Pierre and Marie Curie, Paris, France

Artificial vision heavily relies on perspective cameras because of the constant increase in resolution at low cost due to the mass production. However, a perspective camera has many limitations in autonomous navigation where large fields of views are preferred over large resolutions. An alternative to overcome some of these limitations has been envisioned with the non-central vision model. This chapter gives an overview of what non-central sensors are formally and how they can be extended beyond omnidirectional vision sensing. Finally, since non-central vision has been observed in many reactive visual navigating insects, a solution inspired from them is shown as proof of concept in robotic application.

5.1. Introduction

"Non-central vision" is a name coined by the omnidirectional vision community that flourished during the 2000s. In contrast to the traditional pinhole, perspective camera that images 3D scene as projections onto a plane, with respect to a single focal point or viewpoint, the non-central vision sensors image scene from more than a unique viewpoint. Etymologically, the distinction between central and non-central projection arose from studies on catadioptric devices that show how imaging properties, when scene reflected by a mirror into the camera, can change dramatically according to the alignment of the sensor and the reflector. Maintaining a unique center of projection all along the ray path requires a highly precise placement of the components, which makes the whole device inflexible. Schematics in Figure 5.1 show how the two projections are achieved geometrically. Most of the time, non-perspective images are obtained from catadioptric sensors and they are not practical to deal with when only standard image processing techniques that assume perspective projection are available. However, there are good reasons for the computer vision community to steer the research toward these imaging systems due to:

– the omnidirectional view that allows us to perceive the environment more efficiently;

– the wider imaging possibilities that lift many perspective cameras' limitations.

These properties are promising a faster visual perception for machines and opening the way to low latency autonomous systems with a limited energy budget.

Figure 5.1. *Central and non-central projection*

In robotic vision, classical perspective cameras have a typical field of view (FoV) ranging from $30°$ to $60°$. Beyond these angles, geometric distortions are usually substantial, removing the linear properties of the camera. Geometric corrections exist to compensate the distortions but add a layer of computation not always welcomed. However, for navigation purposes, the benefit of a large FoV exceeds the loss of spatial resolution and the cost for correcting distortions (Streckel and Koch 2005). At the same time, biological evidence has shown that insects and arthropods have via evolution chosen wide FoV over spatial resolution for visual navigation efficiency (Srinivasan et al. 1997). For these reasons, omnidirectional vision sensors are built either by using $180°$ FoV fisheye lenses, which are expensive to manufacture, or by building catadioptric sensors that consist of a combination of a perspective camera and a reflector/mirror. Whatever the chosen solution is, we are facing two families of geometry: the central and the non-central projections.

The central projection geometry is an important model on top of which all the classical computer vision is built. For this reason, central vision sensors form an important sensor family. As these are addressed with more detail in other chapters, we are only providing here a short summarization: the central projection geometric model has a unified formulation proposed by Geyer and Daniilidis (2000) that encompasses the projection mechanism of a perspective camera and the projection mechanism of a central catadioptric vision sensors (i.e. using paraboloid, ellipsoid or hyperboloid reflector).

A lot of our attention has been focused on the central vision model because it allows the preservation of perspective geometry properties, however it is also a strong spatial constraint set on the components. This is where non-central projection models are needed to handle cases where the central projection is not met, either "by accident" or on purpose. The objective of this chapter is to give an overview of the non-central vision geometry and its use which extends far beyond catadioptric omnidirectional vision sensing.

5.1.1. *Generalities*

State-of-the-art omnidirectional vision was established a decade ago, and there have been no major changes since then. The unified central projection model is predominating, meaning that practically used omnidirectional sensors are exclusively central ones, that is, a paraboloid mirror coupled with a camera in most cases, however fisheye lenses are also a preferred choice as they are more compact and steadier than catadioptric devices. Non-central vision sensors have a more interesting story to tell.

One task to carry out before processing the omnidirectional images is to establish the mapping of a 3D point to the pixels in the image. This is similar to the camera calibration operation with the additional reflector inserted between the sensor and the scene. The second task is to unfold the image from the reflector induced distortions to restore perspective geometry. As we will see, these are the main objectives in dealing with catadioptric omnidirectional sensors.

The design of a non-central catadioptric sensor achieving some specific imaging properties is not straightforward. The techniques that can be found in the literature are based on solving differential equations and can be classified into two classes: the ones based on the hypothesis of a reflector with an axis of symmetry that simplifies the mirror shape parameterization into a mono-dimensional function representing the reflector "profile"; the others are formulated in a more general form of a 2D surface to estimate with the assumption that a sample of normal to the surface to recover is provided.

Design and calibration of catadioptric sensors are actually intimately entangled problems because the way light rays are bent into the image plane is a direct result of the reflector's geometry. If the reflector has been designed from a prior process, the calibration procedure usually implies the recovery of its properties such as the caustic surface and/or the sets of normals defining how local the reflector is.

5.1.2. *Biological eyes*

As the initial sources of inspiration for omnidirectional machine vision, insects eyes also offer a large variety of non-central vision. They beat all man-made imaging systems in terms of anteriority, variability, originality and most importantly efficiency. One of the most striking examples is the largely represented family of compound eyes adopted by the insect kingdom. Compound eyes, also called faceted eyes, are famous among non-entomologists because of how efficiently insects navigate: who has never had frustrating experiences of swatting mosquitoes? How can such small beings manage to fly and dodge swatting so quickly? Roboticists are interested in reproducing such efficient visual perception on machines, efficient both in energy consumption and low latency.

Compound eyes are classified into two families: the appositional and the superpositional compound eyes. Though they share similar aspects from the outside (almost hemispheric and multifaceted structure), the internal organization is different as shown in Figure 5.2. Each facet, also called ommatidium, has a narrow FoV that samples light from a selective incident direction; however, the spherical surface tiled by the facets is providing an omnidirectional view of the environment. The appositional eyes structure is best depicted by the stereotypical mosaic of slightly shifted images since each ommatidium is redirecting an incident light into a single rhabdoms, the structure that houses the photosensitive proteins that process light. Such a structure is found in diurnal insects and seems to confer optimal properties for motion perception as it will be elaborated in incoming sections. The superposition eyes structure is radically different as each ommatidium is connected to several rhabdom, leading to a superposition of the signal to produce a single integrated image in the neuron. The achieved light sensitivity is usually orders of magnitude larger than the appositional structure sensitivity. As a result of evolution, this latest architecture is favored by nocturnal insects (Warrant 2017).

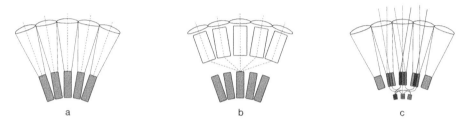

Figure 5.2. *(a) Appositional, (b) superpositional and (c) neural superpositional compound eyes internal structure. For a color version of this figure, see www.iste.co.uk/vasseur/omnidirectional.zip*

As one can see, the reason to include the compound eyes in this chapter's narrative is that they form a family of non-central vision sensors where each facet or cluster of facets are mapped to a viewpoint. While the biological compound eyed vision is not yet fully understood, inspiration from insects that efficiently navigate with it, despite their small brain, can help us to design a more efficient vision sensing for robot navigation. The plenoptic sensors are the artificial systems that loosely reproduce the compound eyes properties and were applied in the early 2010s for autonomous navigation.

5.2. Catadioptric sensors: reflector computation

The non-central sensor is a form of computational imaging sensor if the optic bending the light from the scene to the camera is designed to achieve specific projection into the camera. Here, we are presenting two practical cases where mirrors in a catadioptric system can be calculated to produce almost perspective images or mirrors that preserve angular or spatial resolution in the focal plane. Most of them will be estimated numerically as solutions to partial differential equations relating the scene to the wanted images. In a more general case, the concept of caustic will generalize the definition of viewpoints in a non-central vision sensor and single viewpoint system is a particular case of it. The principles of caustic imaging are then derived from it.

5.2.1. *Caustic surface of a catadioptric system*

So far, the distinction between central and non-central systems is reduced to the question of whether a scene can be imaged perspectively from the world to the mirror and from the mirror to the camera. This problem is equivalent to knowing whether the catadioptric device has a global single viewpoint. However, this is disregarding a family of optical imaging systems rich in variety and all the possibilities they offer. As we are going to focus on what geometrically and optically the non-central status implies, we will also introduce the definition of the caustic surface of an optical system.

The Latin etymological meaning for "caustic" is "burning". In an optical system, the caustic surfaces are the patterns of light that can be seen when light rays go through or are reflected by the optical medium. These patterns are locations where light energy is focused, hence they can burn in the exact way a magnifying lens is used to light a fire. A more formal definition of the caustic surface is the envelope of rays reflected (respectively, refracted) by a surface (respectively, by a material the light passes through). Each point on the caustic surface is a tangent point to the rays and each of them can be seen as a viewpoint of the non-central system.

As the caustic is the locus of the viewpoints, the shaping of the mirror has a direct impact on it and the imaging properties. But the caustic serves an even more important purpose in machine vision: the catadioptric vision sensor calibration. This operation can be summarized by finding the mapping between the viewing direction in the scene and the image point in the camera. This mapping is mandatory for any further metrology considerations. In short, computing the caustic surface is calibrating the entire catadioptric vision sensor.

There are mainly two ways to proceed in computing the caustic surface: one is derived from geometrical optic considerations for which the surface, as the envelope of rays, "is the loci of singularities in the flux density" (Burkhard and Shealy 1973). This approach can lead to the closed-form expression of the caustic surface for the central catadioptric sensor or a numerical approximated form for some more general mirror shape. The second technique is based only on geometry where the mirror can be approximated locally by an ellipse under the assumption of axial symmetry. The following sections give guidelines for the two techniques.

5.2.2. *Caustic surface computation*

5.2.2.1. *Geometrical optic*

This technique is based on the analysis of the flux density of the light beam reflected by a deflector into a receiver as detailed in Burkhard and Shealy (1973): if dS_1 and dS_2 are the respective area elements on the deflector (mirror) and on the receiver (sensor), the caustic is then where in space the ratio $\frac{dS_1}{dS_2}$ vanishes (see Figure 5.3). This ratio is equivalent to the Jacobian defined by the reflector coordinate frame and the receiver coordinate frame. The technique expressed the incident ray as a function of the normal to the reflector $\mathbf{n}$ and the reflected ray $\mathbf{i}$

$$\mathbf{i} = \mathbf{r} - 2\mathbf{r}^T\mathbf{n}\mathbf{n}, \qquad\qquad [5.1]$$

and the incident beam is a line parameterized by a point it passes through, $\mathbf{F}$, with a unit incident direction vector, $\mathbf{i}$ and real scalar t such that

$$\mathbf{I} = \mathbf{F} + t\mathbf{i}. \qquad\qquad [5.2]$$

A usual assumption made on the reflector is that it has a rotational symmetry. This makes the parameterization simpler with one single parameter u such that $\mathbf{F} = \left(F_x(u),\ F_y(u)\right)^T$. Then the Jacobian of $\mathbf{I}$ is given as

$$J(\mathbf{I}) = \det \begin{pmatrix} \frac{\partial I_x}{\partial u} & \frac{\partial I_x}{\partial t} \\ \frac{\partial I_y}{\partial u} & \frac{\partial I_y}{\partial t} \end{pmatrix}. \qquad\qquad [5.3]$$

The caustic point belongs to the line $\mathbf{I}$ and is located for t such that $J(\mathbf{I}) = 0$, that is, when t satisfies

$$t = \frac{\frac{\partial F_x}{\partial u}i_x - \frac{\partial F_y}{\partial u}i_y}{\frac{\partial i_x}{\partial u}i_x - \frac{\partial i_y}{\partial u}i_y}. \qquad\qquad [5.4]$$

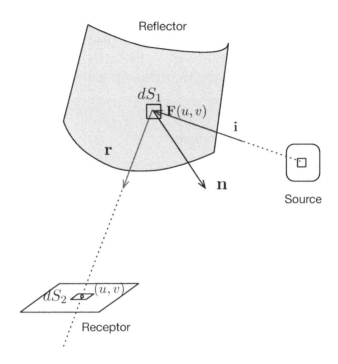

Figure 5.3. *Caustic for a catadioptric device made of a source,*
a reflector and a receptor. For a color version of this figure,
see www.iste.co.uk/vasseur/omnidirectional.zip

This technique is not restricted to reflector with an axis of symmetry. The Jacobian of $\mathbf{I}$ and the expression for t are more complex if the symmetry constraint is lifted, but the concept is similar. And when reflector does not have an analytical form, local interpolation such as spline allows for estimating the Jacobian numerically. This is the procedure explored by Swaminathan et al. (2001) for a large variety of reflector shapes.

5.2.2.2. *Geometrical approximation*

The geometric variation is an alternative to cases where mirror profile/shape is not closed-form defined, for example, given as a point cloud. This approach, as presented in Ieng and Benosman (2006), generalizes the local approximation method introduced in Bruce et al. (1981) for 3D mirrors with axial symmetry. The idea is to locally find an ellipse that locally fits the reflector and use the property that the associated caustic point is one of the foci of the ellipse (see Figure 5.4).

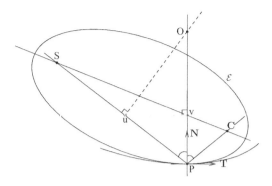

Figure 5.4. *Geometric construction. For a color version of this figure, see www.iste.co.uk/vasseur/omnidirectional.zip*

If we consider the reflector parameterization: $\mathbf{F} = (X, \ F(X))^T$, then the property established in Bruce et al. (1981) is stated as follows.

Given two regular curves F and G of class C^n, with a common tangent at a common point P, taken as $(0,0)^T$ and the x-axis as the tangent. Then P is an n-order point of contact if

$$
\begin{cases}
F^{(k)}(0) = G^{(k)}(0) = 0 & \text{if } 0 \le k < 2 \\
F^{(k)}(0) = G^{(k)}(0) & \text{if } 2 \le k \le n - 1 \\
F^n(0) \ne G^{(n)}(0)
\end{cases}
\tag{5.5}
$$

There is only one conic $\mathcal{E}$ of at least a 3-point contact with F at P, where S and C are the foci. C is the caustic point of F at P, with respect to S.

With $\mathbf{P} = (x_p, F(x_p))^T \in \mathbf{F}$, the curvature of $\mathbf{F}$ at $\mathbf{P}$ is equal to

$$
k = \frac{F''(x_p)}{\sqrt{1 + F'(x_p)^2}^3}.
\tag{5.6}
$$

From here, we can build $\mathbf{C}$, step by step:

– compute $\mathbf{O}$, the center of curvature of F at $\mathbf{P}$ as: $\mathbf{O} = \mathbf{P} + \frac{1}{|k|}\mathbf{N}$;

– project orthogonally $\mathbf{O}$ to (SP) at $\mathbf{u}$; project orthogonally $\mathbf{u}$ to (PO) at $\mathbf{v}$. (Sv) is the principal axis of $\mathcal{E}$;

– set $\mathbf{C}$ on (Sv) such that (OP) is bisecting the angle $\widehat{SPC}$.

This local fitting of a planar ellipse is valid as long as the incident, reflected rays, and the normal to the surface at the incident point are all coplanar. This is only true for a mirror with a rotational symmetry axis. Planes intersecting the mirror that contain the symmetry axis are satisfying such a requirement and the computation of the caustic surface is done by applying the local fitting of ellipses by sampling the reflector into such planes.

As the caustic is now computed point by point from each incident ray to reflector, the calibration of the whole sensor is straightforward if the pose of the perspective camera with respect to the reflector can be estimated. This step is often achieved by detecting no coplanar edges of the reflector (Fabrizio et al. 2002; Agrawal et al. 2010) or by some equivalent method.

5.2.3. *Reflector computation*

5.2.3.1. *Differential approach for reflector with axial symmetry*

Catadioptric non-central sensors form a large subset among omnidirectional sensors where space constraint has been relaxed to achieve specific imaging geometries. They are built similarly as a combination of a perspective camera and a reflector which, for manufacturing reasons, are designed as revolution surfaces such that the geometry properties are functions of the distance to the revolution axis. Due to the symmetry, it is often sufficient to know a cross-section (or "profile") of the mirror to calculate which pixel an incident ray is imaged into. This cross-section is referred to as a function F of X, where X is the distance of the 3D point on the mirror to the axis (see Figure 5.5). Early works (Hicks and Bajcsy 2000; Gaspar et al. 2002) proposed techniques to map incident rays into desired pixels to minimize distortions and restore perspective geometry as much as it is possible. To relieve camera placement constraints, it is customary to use a telecentric lens to reject the camera viewpoint to infinity. With such hypotheses, the incident and the reflected angles at F(X) on the profile satisfy the following equation:

$$\phi = \theta + 2 \arctan F'(X). \qquad [5.7]$$

θ is also referred to as the direction incident to the effective viewpoint c and $F'(X)$, the slope of the tangent to F at X. This mapping is generic as it does not assume any prior hypothesis on F and imposes nothing about where the camera must be placed with respect to the reflector but its lens optical axis to be collinear with the reflector's axis of revolution. A camera calibrated with respect to the reflector provides the explicit mapping of θ to the pixels and conversely.

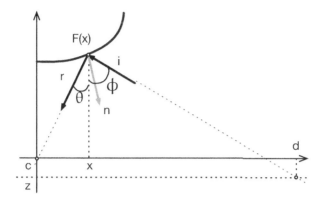

Figure 5.5. *Incident ray imaged into a the camera after reflection from mirror F. The incidence and reflection can be parameterized by angle ϕ. For a color version of this figure, see www.iste.co.uk/vasseur/omnidirectional.zip*

From Figure 5.5 and equation [5.7], classical geometry provides the following three additional equations:

$$
\left\{
\begin{array}{rl}
\frac{X}{F} & = \tan\theta \\
\frac{d-X}{F(X)} & = \tan\phi \\
\phi & = \theta + 2\arctan F(X)'
\end{array}
\right. ,
\qquad [5.8]
$$

where $(z, d)^T$ is the 3D scene point, source of the incident ray **i** to the mirror at $(X, F)^T$ and reflected to the camera at **c** in the direction of **r**. The combination of the three equations leads to the differential equation

$$
\frac{\frac{X}{F} + \frac{2F'}{1-F'^2}}{1 - \frac{X}{F}\frac{2F'}{1-F'^2}} = \frac{d-X}{F-z},
\qquad [5.9]
$$

equivalent to

$$
F'^2 + 2\frac{(F-z)F + (d-X)X}{(z-F)X + (d-X)F}F' - 1 = 0.
\qquad [5.10]
$$

F' is known as the shaping function in the literature.

Solving this differential equation – mostly numerically due to its nonlinearity – for $(F, X)^T$ with known $(z, d)^T$ allows for the design of a mirror with specific geometrical imaging properties. In Hicks and Bajcsy (2000), a mirror achieving an approximated perspective projection between planes parallel to the floor, within a range of height, is computed using this technique. The mirror is also constrained to map orthographically the floor into the sensor focal plane. With this additional constraint, the shaping function is simplified into

$$F'^2 + 2\frac{F}{d - X}F' - 1 = 0. \hspace{3cm} [5.11]$$

The exact perspective projection between all planes parallel to the floor is not possible as mentioned by the authors precisely because of the non-central projection achieved by the mirror. Gaspar et al. (2002) used the shaping function from equation [5.10] to inject imaging constraints to get the desired mirror, for example, if an equi-resolution projection is required, we have to set d (respectively z) as a linear function of $\frac{X}{F}$, that is, $d = a\frac{X}{F} + b$ and z (respectively, d) as a constant: this leads to either a function mapping a vertical (respectively horizontal) distance of the scene into a constant distance in the image plane. However, there is no guarantee of a solution to the differential equation.

5.2.4. *Methods for reflector with no axial symmetry*

5.2.4.1. *Spline interpolation approaches*

These approaches assume the user to provide a set of scene points $\{\mathbf{M}_k\}$, mapped to a set of corresponding reflected ray directions $\{\mathbf{r}_k\}$ by the reflector surface at $\{\mathbf{F}_k\}$. The $\{(u_k, v_k)^t\}$ are pixels on the imager. In general, the mapping of the $\{\mathbf{r}_k\}$ to the pixels is related to the optic used in the imager, for example, a projection matrix for a standard perspective camera. In such a case, $\mathbf{r}_k$ is colinear to $\mathbf{m}_k = \widetilde{P}(u_k, v_k, 1)^T$, where $\widetilde{P}$ is the pseudoinverse of the projection matrix. The mirror surface is estimated iteratively by an optimization algorithm from the equalities satisfied by the incident and reflected rays and the normals to mirror

$$\begin{cases} \mathbf{i}_k &= \frac{\mathbf{M_k}-\mathbf{F}_k}{||\mathbf{M_k}-\mathbf{F}_k||} \\ \mathbf{r}_k &= \frac{\mathbf{m}_k-\mathbf{F}_k}{||\mathbf{m}_k-\mathbf{F}_k||} \end{cases}. \hspace{3cm} [5.12]$$

We have two additional scalar equalities satisfied by the normal to the reflector at $\mathbf{F}_k$ since the partial derivatives of $\mathbf{F}$ are also tangent to the mirror at $\mathbf{F}_k$

$$\begin{cases} \frac{\partial \mathbf{F}_k}{\partial u}^T \mathbf{n}_k = 0 \\ \frac{\partial \mathbf{F}_k}{\partial v}^T \mathbf{n}_k = 0 \end{cases},$$ [5.13]

with $\mathbf{n}_k$ being the normal at $\mathbf{F}_k$. Finally, the following equation relates $\mathbf{F}$ to the reflected directions

$$\mathbf{F}_k = \mathbf{C}_k + D(u, v)\mathbf{r}_k,$$ [5.14]

where $\mathbf{C}_k$ is the so-called viewpoint as it is introduced in Swaminathan et al. (2003) alongside with [5.14] and D is a 2D spline function to approximate locally the distances of $\mathbf{C}_k$ to the $\mathbf{F}_k$

$$D(u, v) = \sum_{i,j} c_{ij} f_i(u) g_j(v).$$ [5.15]

The combination of [5.13] and [5.14] yields

$$\begin{cases} \frac{\partial (\mathbf{C}_k + D(u,v)\mathbf{r}_k)}{\partial u}^T \mathbf{n}_k = 0 \\ \frac{\partial (\mathbf{C}_k + D(u,v)\mathbf{r}_k)}{\partial v}^T \mathbf{n}_k = 0 \end{cases} \Rightarrow \begin{cases} -\frac{\partial \mathbf{C}_k}{\partial u}^T \mathbf{n}_k = \left(\frac{\partial D}{\partial u}\mathbf{r}_k + D\frac{\partial \mathbf{r}_k}{\partial u}\right)^T \mathbf{n}_k \\ -\frac{\partial \mathbf{C}_k}{\partial v}^T \mathbf{n}_k = \left(\frac{\partial D}{\partial v}\mathbf{r}_k + D\frac{\partial \mathbf{r}_k}{\partial v}\right)^T \mathbf{n}_k \end{cases}$$ [5.16]

To solve [5.16] for D, in the more general use of the method proposed in Swaminathan et al. (2003), users/designers are requested to initialize a set of facets on a plane, placed at a chosen distance from the scene and sensor. The initial set of normals at these facets is calculated from the incidents and reflected vectors as given in [5.12]

$$\mathbf{n}_k = \frac{\mathbf{r}_k - \mathbf{i}_k}{||\mathbf{r}_k - \mathbf{i}_k||}$$ [5.17]

Equation [5.16] is to be solved in D and it can be rewritten into a linear form of $A\mathbf{c} = \mathbf{b}$, where $\mathbf{c}$ contains the coefficients c_{ij} of D. The mirror shape can then be estimated with an iterative, least squared error minimization algorithm such as a gradient descent.

5.2.4.2. *Gradient approximation*

The mirror shape estimation presented in Kiser and Pauly (2012) approaches the problem in the inverted way: the image sensor is replaced by a light source and the scene is a surface on which image points are mapped adequately with a mirror. This interpretation is equivalent to the previous one as optical paths are invertible in optical geometry. The method is also more constrained as incident (respectively, reflected) rays are assumed to be parallel as illustrated by Figure 5.6.

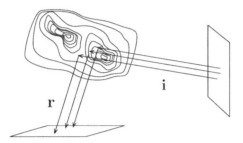

Figure 5.6. *Caustic surface with source at infinity and incident plane at infinity. For a color version of this figure, see www.iste.co.uk/vasseur/omnidirectional.zip*

The mirror shape is parameterized in the form of $\mathbf{F} = (x, y, z(x,y))^T$. This allows for expressing the normal to the mirror as a collinear vector to the cross product of the partial derivatives of $\mathbf{F}$

$$\mathbf{n} \equiv \frac{\partial \mathbf{F}}{\partial x} \times \frac{\partial \mathbf{F}}{\partial y} = \left(-\frac{\partial z}{\partial x}, -\frac{\partial z}{\partial y}, 1 \right)^T \Rightarrow \begin{cases} -\frac{\partial z}{\partial x}(x,y) = \frac{n_x}{n_z} \\[2mm] -\frac{\partial z}{\partial y}(x,y) = \frac{n_y}{n_z} \end{cases} \qquad [5.18]$$

This equation is made into a linear form of $Ax \sim b$, with A being the operator for the finite differences approximating the spatial derivatives of z, x the vector of values $z(x,y)$ sampled over a regular grid and b contains the components of the normal vectors to $\mathbf{F}$ at each of the samples.

In Kiser and Pauly (2012), from samples obtained over a $N_x \times N_y$ grid, with N_x and N_y not smaller than 3, the vector x is built by listing samples row-by-row starting from the first column to the last. Due to this ordering, A is shaped as

$$A = \begin{bmatrix} A_x \\ A_y \end{bmatrix} ; A_x = D_{N_x} \otimes I_{N_y} ; A_y = I_{N_x} \otimes D_{N_y}, \qquad [5.19]$$

where $\otimes$ is the Kronecker product and

$$
D_N = \begin{bmatrix} -1 & 1 & & \cdots & \\ & -1 & 1 & \cdots & \\ & & \ddots & & \\ & & & -1 & 1 \end{bmatrix}
$$

is a $(N-1) \times N$ matrix.

Solving [5.18] for x, given the normals n can be achieved by a least square minimization technique. (Kiser and Pauly 2012) suggest applying the Gauss–Seidel or gradient descent algorithms for that purpose.

5.2.4.3. *Closure*

Both techniques are unsurprisingly similar as they require prior knowledge of the incident and reflected directions, which lead to identifying the local normals to the reflector. Then, both find the solution by solving a first-order differential equation relating the set of normals to the reflector, either by fitting a differentiable function (e.g. a spline function) or by approximating the gradient calculation. The main difference between the two techniques is given in Kiser and Pauly (2012): the incident and reflected directions are global, that is to say the incident rays form a parallel bundle and the reflected rays form another parallel bundle. This is equivalent to constraining the viewpoints and the source at infinity. In Swaminathan et al. (2003), these directions are not subject to such limitations and can be locally different, hence this technique is more generic.

5.3. Plenoptic vision as a unique form of non-central vision

While works on omnidirectional vision made a clear dichotomy between central and non-central sensors, the concept of a single viewpoint in visual sensing is not new and predates much earlier before the omnidirectional vision became a research topic in computer vision: binocular stereo-vision is the most obvious and simplest example and sensors built from sets of independent cameras are a straightforward extension of the principle of multiple views. One early attempt of interpretation was formulated by da Vinci as "infinite visual pyramids" emitted by an object. These "pyramids" are the cones defining the FoV of the perspective cameras placed at the vertices of the cones. A formalization has been proposed in Adelson and Bergen (1991), and the authors designed the "plenoptic function" that captures the complete information of the optical signal. This function provides the intensity of each light ray as a function of the wavelength, time, viewpoint and viewing directions. Chronologically, the physical implementation of the plenoptic camera has been built in 1908 (Lippmann 1908) before the theory behind it had been established. Modern designs of the plenoptic cameras came later, first with descriptive analyses derived from the plenoptic function (Adelson and Wang 1992; Neumann et al. 2002), then prototypes used in robot navigation (Dansereau et al. 2011; Dong et al. 2013) and high-end compact imaging devices have been commercialized as large consumer products from Lytro and Raytrix. This section provides a summarization of the mathematical model of the plenoptic camera and a robotic application and readers may refer to Dong (2012) for complementary details.

5.3.1. *Formalism and design*

5.3.1.1. *The plenoptic function*

The plenoptic function is meant to be a complete physics model of light that describes an image from a bundle of sampled rays where each ray is characterized by the position and orientation from which it is seen at a given time t. In a large case of application, the function is simplified by dropping the dependence on the wavelength as visible light is assumed by default. The simplified plenoptic function is defined as $L(\mathbf{x}, \mathbf{r}, t)$, with:

- L being the radiance function;
- $\mathbf{x} = (x, y, z)^T$ the viewpoint in space;
- $\mathbf{r} = (\theta, \phi)^T$ the direction in polar coordinate from which the L is seen;
- t being the time.

A more practical parameterization for L, with simplified hypotheses, was proposed by Levoy and Hanrahan (1996). The authors assume the plenoptic function being sampled by a set of perfectly identical perspective cameras placed over a regular planar grid. This consists of using two parallel planes to encode simultaneously ray directions and viewpoints positions. The first plane is defined by all the viewpoints and the second is defined by the image planes. Any incident ray is then represented by the pair viewpoint $\mathbf{C}$ and the image point $\mathbf{m}$ such as (see Figure 5.7):

$$- \mathbf{C} = (x, y, 0)^T;$$
$$- \mathbf{m} = (u, v, f)^T;$$

where f is the cameras' focal length. This parameterization straightforwardly encodes viewpoints and viewing directions as required for the plenoptic function, which is then defined as $L(x, y, u, v, t)$. This two-planes parameterization is suggesting how actual physical plenoptic cameras, for practical reasons (i.e. calibration, synchronization and so on), will be built as explained in the next section.

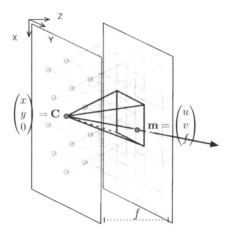

Figure 5.7. *Two-planes parameterization to encode rays by intersections of incident rays with the image plane and the viewpoints plane. For a color version of this figure, see www.iste.co.uk/vasseur/omnidirectional.zip*

5.3.2. *Plenoptic camera*

The typical embodiment of the plenoptic function is the "plenoptic camera" as studied in Adelson and Wang (1992), but was physically implemented early on in the last century by Lippmann (1908). The idea behind the plenoptic camera is the regular juxtaposition of smaller perspective cameras whose viewpoints are regularly spanning a common plane, forming the locus of viewpoints (i.e. the caustic surface as introduced in the previous section) and with the assumption of having the same optical properties (e.g. same focal lengths, same FoV and so on). Under this assumption, all these cameras' image planes are also coplanar and parallel to the locus of viewpoints. Hence, calibration and modeling of such a sensor are made easier.

The work presented in Adelson and Wang (1992) seems to be the first in formalizing the computational sensing properties of the plenoptic camera by describing how depth can be estimated with a single lens device. However, it is not strictly speaking a single lens system, but rather a sensor made of a principal lens that focuses light to an array of microlenses that then are focusing it on the sensor itself. Depths are estimated by estimating the disparity via a correlation-like operation.

This principle of using microlenses arrays to split the sensor into several smaller perspective cameras is the chosen design for compact plenoptic vision sensors, also referred to as light field cameras. The trade-off for compactness is a reduced depth sensitivity since stereo-bases, for example, distances between viewpoints are constrained due to integration.

5.3.3. *Applications in robotic navigation: plenoptic visual odometry*

The plenoptic vision sensor has many promising computing properties, however years of attempts in putting them into large consumer use are unconvincing. It seems to be useful for machine vision rather than for human interaction/application purposes. In embedded mobile devices, such a sensor is mainly aiming for after-shot refocusing to produce better images. From this perspective, such a computational imaging device looks more like a gadget.

On the other hand, autonomous robotics are gaining a better vision sensor for vision-based navigation. We will give, in the following section, a formal explanation on how to derive motion from the plenoptic function since it also encodes temporal information and some actual experiments of using plenoptic cameras to navigate.

Motion can be recovered from the plenoptic function by applying the method presented in Neumann et al. (2002). The vision sensor is moving through the scene and each of its cameras at x samples the plenoptic function from direction r (expressed in spherical coordinate) at time t. The ego-motion is rigid, that is, given by a rotation matrix R and a translation vector $\mathbf{T}$.

Estimating the motion parameters from the plenoptic function assumes the smoothness of the signal, which allows Taylor's first-order expansion. A second hypothesis is a constant scene illumination over time. If this is fulfilled, the plenoptic function satisfies the photo-consistency constraint which is similar to that used in optical flow computation

$$-L_t = \nabla_x L^T \frac{d\mathbf{x}}{dt} + \nabla_r L^T \frac{\mathbf{r}}{dt}, \qquad [5.20]$$

where $\nabla_x L$ and $\nabla_r L$ are the spatial gradient of L and L_t, the partial temporal derivative. If we introduce $\mathbf{q}$ and $\boldsymbol{\omega}$ as, respectively, the linear and angular velocities, the previous equation is equivalent to

$$-L_t = \nabla_x L^T \mathbf{q} + (\mathbf{x} \times \nabla_x L + \mathbf{r} \times \nabla_r L)^T \boldsymbol{\omega}, \qquad [5.21]$$

where $\times$ is the cross-product operator.

Dealing with Cartesian and spherical coordinates in the same equation is not practical, thus this is where the two-planes parameterization comes into play with an adequate coordinate change. Equation [5.21] can be rewritten into the equivalent form of

$$-L_t = \begin{pmatrix} L_x & L_y & L_u & L_v \end{pmatrix} \begin{pmatrix} 1 & 0 & \frac{-u}{f} & -\frac{uy}{f} & \frac{ux}{f} & -y \\ 0 & 1 & \frac{-v}{f} & -\frac{vy}{f} & \frac{vx}{f} & x \\ 0 & 0 & 0 & -\frac{uv}{f} & \frac{u^2}{f}+f & -v \\ 0 & 0 & 0 & -\frac{v^2}{f}-f & \frac{uv}{f} & u \end{pmatrix} \begin{pmatrix} \mathbf{q} \\ \boldsymbol{\omega} \end{pmatrix}. \qquad [5.22]$$

Solving this linear equation for the motions parameters can be done with regular least-square minimization techniques, assuming we have sufficient measurements to build an over-determined equation system. The core idea of the plenoptic visual odometry has been applied to a simulated scenario based on real data acquired by an unmanned underwater vehicle in Dansereau et al. (2011): the plenoptic sensor in this work is virtual/simulated. This technique has been then applied to the navigation of an autonomous vehicle in Dong et al. (2013) that embeds a physical array of 3×3 synchronized cameras. This latest work improved the approach by adding a multi-scale representation of the plenoptic function.

5.3.3.1. *Application to indoor navigation*

This section is showing some samples of results from Dong (2012). The plenoptic vision sensor is built out of a set of nine perspective cameras arranged in a 3×3 array. These cameras are configured to work synchronously and with the same frame rate. The cameras' interface allows high-speed data transfer and the image maximal resolution is set to 320×240 pixels as a trade-off between signal resolution and available bandwidth allowed by the connection to the computer. A rigid mount was designed to ensure the stability of the array structure and it fixes the spacing of the cameras at 5 cm both horizontally and vertically. The cameras' synchronization is achieved by sending simultaneously an external triggering signal to start the acquisition from a micro-controller, MSP430 LaunchPad. The triggering signal is a 20 Hz impulse train, generated by the embedded system development tool from the IAR system (IAR-EW-430), so the frame rate of the cameras is set to its maximal value of 20 fps. The entire visual system is embedded on a mobile Pioneer platform to record data, while the latter executes trajectories at various speeds up to 1.8 m/s.

The experiments are monitored by the motion capture system from Vicon. The Vicon system tracks the robot/camera array at high frequency (in this case, we set its frequency as 100 Hz which is five times the cameras' frequency), and it delivers 6 degrees-of-freedom motion with high positional and angular accuracy. A systematic accuracy analysis is done in Windolf et al. (2008), where the authors claim that the Vicon system provides an overall accuracy of 63 ± 5 μm. Figure 5.8 shows four estimated trajectories from six trials for each. Each of them is estimated with the plenoptic visual odometry for a total of 100 samples; only a subset of them are plotted in red against the Vicon ground truth in blue for readability reasons. Figure 5.9 shows the histogram of the translational errors for all the mentioned motions. These errors are represented as absolute errors with a mean value equal to 14 cm. All trajectories have an average length of 20 m.

5.3.3.2. *Closure*

As the embodiment of the plenoptic function designed to encode "the complete holographic representation of the visual world" (to quote Adelson and Bergen (1991)), the plenoptic sensor is the most suitable in relating the sensor's motions to the visual inputs. And since motions and 3D structures are closely related, the path to localization and mapping from visual inputs is obvious as it was shown here with practical robotic examples.

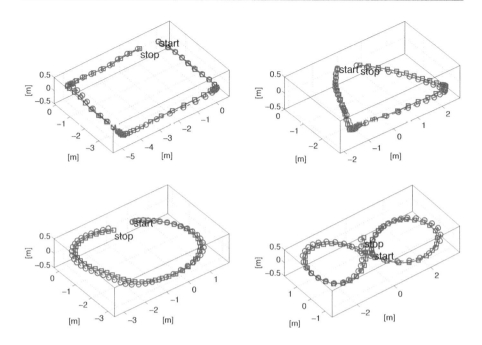

Figure 5.8. *Set of trajectories built from the plenoptic estimation (red squares), compared to the ground truth (blue squares) from the Vicon tracking in a working volume of 5 × 5 × 5m³. The average velocity is 1.5 m/s. For a color version of this figure, see www.iste.co.uk/vasseur/omnidirectional.zip*

The introduction of the plenoptic sensors, also called light field cameras, to a broader audience was attempted by several companies that made plenoptic sensors with microlenses arrays, with Lytro and Raytrix being the most well known. Due to the small stereo basis of these compact devices, applications are constrained to close range (<1 m) observations (Rangappa et al. 2019). Lytro bet on the consumer market but failed and has shut down operations since 2018, whereas Raytrix (2022), which has focused on the more specific field of scientific vision sensor, is still operating.

5.4. Conclusion

The non-central projection is a generic model for the geometry of imaging in vision sensors. The principles subtended by the non-central projection have been studied a century ago when image pioneers were exploring multiple views geometry. This model is useful in comprehending how to build sensors with specific and handcrafted imaging properties, how to calibrate complex and nonlinear vision sensors and how to estimate the complex stereo-vision mechanisms that relate structures to sensors' motion.

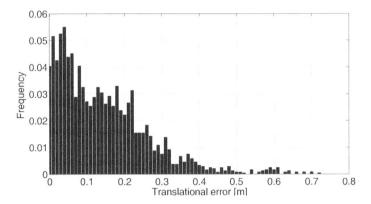

Figure 5.9. *Histogram of the translational errors. The average length of the trajectories is 20 m. For all trials, the mean translational error is 0.14 m with a standard deviation of 0.12 m. For a color version of this figure, see www.iste.co.uk/vasseur/ omnidirectional.zip*

A complex and nonlinear geometry may imply additional and costly preprocessing of the visual inputs with standard computer vision approaches. The biological compound eyes of some insects, however, seem to have solved the problem efficiently with little power consumed, allowing for a low latency perception–action loop. Insects' non-central vision perception is a good source of inspiration for machine vision used in autonomous navigation where energy and low latency are major constraints.

Sensors are not the main and only causes of performances in visual perception, the intelligence implementing the processing is another bottleneck. The current artificial intelligence (AI) is often assimilated to the concept of deep artificial neural networks achieving impressive image learning tasks such as automated segmentation, faces recognition, or even more exotic and complicated tasks. Such AI consumes a sheer amount of resources (energy, mass storage, etc.) for its training.

To reduce the power consumption, the development of the entire processing pipeline is necessary and this includes redesigning the memory architecture and the processing unit among others. As it is, the standard Von Neumann architecture on which most computing systems are based are not suitable ones, simply because of the separation between the memory and processing units, leading to power-consuming repetitive back and forth data transfers from one to the other. By comparison, the processing of the data itself consumes only a small fraction of the power budget.

5.5. References

Adelson, E.H. and Bergen, J. (1991). The plenoptic function and the elements of early vision. *Computational Models of Visual Processing*, 3–20.

Adelson, E.H. and Wang, J.Y.A. (1992). Single lens stereo with a plenoptic camera. *IEEE Trans. Pattern Anal. Mach. Intell.*, 14(2), 99–106.

Agrawal, A., Taguchi, Y., Ramalingam, S. (2010). Analytical forward projection for axial non-central dioptric and catadioptric cameras. In *Proc. Europ. Conf. Comput. Vis.*, 6313. Springer, Berlin/Heidelberg.

Bruce, J., Giblin, P., Gibson, C. (1981). On caustics of plane curves. *Am. Math. Mon.*, 88, 651–667.

Burkhard, D. and Shealy, D. (1973). Flux density for ray propagation in geometric optics. *J. Opt. Soc. Am. A*, 63(3), 299–304.

Dansereau, D., Mahon, I., Pizarro, O., Williams, S. (2011). Plenoptic flow: Closed-form visual odometry for light field cameras. In *Proc. IEEE/RSJ Int. Conf. Intel. Robots Syst.* IEEE, San Francisco.

Dong, F. (2012). Vision sensor design and evaluation for autonomous navigation. PhD Thesis, University Pierre and Marie Curie, Paris.

Dong, F., Ieng, S., Savatier, X., Etienne-Cummings, R., Benosman, R. (2013). Plenoptic cameras in real-time robotics. *Int. J. Robot. Res.*, 32(2), 206–217.

Fabrizio, J., Tarel, J.-P., Benosman, R. (2002). Calibration of panoramic catadioptric sensors made easier. In *Proc. 3rd Workshop on Omnidirectional Vision*, 45–52. IEEE and ECCV'02, Copenhagen.

Gaspar, J., Decco, C., Okamoto, J., Santos-Victor, J. (2002). Constant resolution omnidirectional cameras. In *Proc. the IEEE Workshop on Omnidirectional Vision*, 27–34. IEEE and ECCV'02, Copenhagen.

Geyer, C. and Daniilidis, K. (2000). A unifying theory for central panoramic systems and practical implications. In *Proc. 6th Eur. Conf. Comput. Vis.*, 445–461. Springer-Verlag, Berlin/Heidelberg.

Hicks, A. and Bajcsy, R. (2000). Catadioptric sensors that approximate wide-angle perspective projections. In *Proc. IEEE Workshop on Omnidirectional Vision (OMNIVIS)*. IEEE, Hilton Head.

Ieng, S.H. and Benosman, R. (2006). Geometric construction of the caustic surface of catadioptric non-central sensors. In *Imaging Beyond the Pinhole Camera*, Daniilidis, K. and Klette, R. (eds). Springer, Dordrecht. doi: 10.1007/978-1-4020-4894-4_3.

Kiser, T. and Pauly, M. (2012). Caustic art. Technical Report, EPFL.

Levoy, M. and Hanrahan, P. (1996). Light field rendering. In *Proc. 23rd Annual Conf. Comp. Graph. Inter. Tech.* ACM, New Orleans.

Lippmann, G. (1908). Épreuves réversibles donnant la sensation du relief. *J. Phys. Theor. Appl.*, 7, 821–825.

Neumann, J., Fernmuller, C., Aloimonos, Y. (2002). Eyes from eyes: New cameras for structure from motion. In *IEEE Workshop on Omnidirectional Vision*. IEEE, Copenhagen.

Rangappa, S., Matharu, R., Petzing, J., Kinnell, P. (2019). Establishing the performance of low-cost Lytro cameras for 3D coordinate geometry measurements. *Mach. Vis. Appl.*, 30(4), 615–627.

Raytrix (2022). 3D light-field machine vision camera [Online]. Available at: https://raytrix.de.

Srinivasan, M., Weber, K., Venkatesh, S. (1997). *From Living Eyes to Seeing Machines*. Oxford University Press, New York.

Streckel, B. and Koch, R. (2005). Lens model selection for visual tracking. In *Proc. 27th DAGM Conf. Pattern Recogn.* Springer, Jena.

Swaminathan, R., Grossberg, M., Nayar, S. (2001). Caustics of catadioptric cameras. In *Proc. IEEE Int. Conf. Comp. Vis.*, 2. IEEE, Vancouver.

Swaminathan, R., Nayar, S., Grossberg, M. (2003). Framework for designing catadioptric projection and imaging systems. Technical Report, Department of Computer Science, Columbia University.

Warrant, E. (2017). The remarkable visual capacities of nocturnal insects: Vision at the limits with small eyes and tiny brains. *Phil. Trans. R. Soc. B*, 372(1717), 20160063.

Windolf, M., Gözen, N., Morlock, M. (2008). Systematic accuracy and precision analysis of video motion capturing systems – Exemplified on the Vicon-460 system. *J. Biomech.*, 41(12), 2776–2780.

6

Localization and Navigation with Omnidirectional Images

Helder Jesus Araújo[1], Pedro Miraldo[2] and
Nathan Crombez[3]

[1]*Institute of Systems and Robotics, University of Coimbra, Portugal*
[2]*Mitsubishi Electric Research Laboratories (MERL), Boston, USA*
[3]*CIAD Laboratory, University of Technology of Belfort-Montbéliard, France*

This chapter describes and critically reviews methods, techniques and theories that address the challenges of visual localization and navigation by taking advantage of the benefits of the omnidirectional vision. In the first instance, the most important configurations of cameras, lenses, mirrors and the commonly used image formation models are recalled through the prism of the intrinsic aspects of this kind of application. Then, the approaches from the literature are detailed with an emphasis on the contributions of the omnidirectional vision. Indeed, increased fields of view, as well as some specific properties of the geometric transformations involved in omnidirectional vision systems are analyzed, specifying their advantages for navigation and localization. Map-based methods, including metric and topological methods are first described. Conversely, mapless approaches that are mainly based on visual odometry are then discussed. Methods aiming to solve both the localization of the omnidirectional camera and the creation of the environment representation problems are then considered. Finally, applications of the multi-robot formation are included, stressing the advantages of this type of vision system.

Omnidirectional Vision,
coordinated by Pascal Vasseur and Fabio Morbidi. © ISTE Ltd 2023.

6.1. Introduction

A wide range of application areas, such as mobile robotics, autonomous transportation and industrial manipulation, just to name a few, require an accurate and effective localization and navigation system. The ability to localize itself and navigate is crucial to ensure the operation of an autonomous mobile robot or vehicle so that their tasks can be fully accomplished. On the one hand, localization can be defined as the estimation of an object pose, that is, its position and its orientation, relative to a reference frame based on data acquired by sensors within a known or an initially unknown environment. On the other hand, navigation consists of the determination of a path between a starting pose and a goal pose for a robot or vehicle traveling between them (Bonin-Font et al. 2008). In general, tasks require specific goals to be reached, which means visiting a specific position in space with a specific orientation, in other words a specific pose. While moving around in their environment, the robots must negotiate obstacles, both static and dynamic, and need to plan their motion according to the structure and map of the environment. Robot navigation entails solving problems of self-localization, map estimation, map interpretation and obstacle detection. Self-localization implies the ability to establish its position within a frame of reference. In contrast, path planning requires the current position, the position of the goal location and the trajectory/path required to reach the goal position to be determined. Localization and navigation are therefore two fundamental joint problems that have been studied for many years and for which numerous solutions have been proposed. The set of localization and navigation capabilities require the extraction of relevant information from the environment using exteroceptive sensors. Exteroceptive sensors allow the extraction of information from the environment, and some of them allow the inference of its physical structure. Sensing options depend on whether the navigation is performed indoors or outdoors. Indeed, depending on the nature of the environment, some types of information are more suitable than others. Many technologies have thus been developed to estimate the pose of an entity based on different kinds of collected information. Consequently, a wide variety of approaches have emerged, using different types of sensors and processing techniques. Consequently, very different approaches have emerged, using different types of sensors, namely GPS, sonar, lidar, radar and vision, and thus different processing techniques.

Many animals can localize and navigate with a relatively poor set of sensors; most animals do not possess sensors that can tell them the distance to the obstacles (Milford 2008). Vision is commonly used by animals to navigate and it is especially useful since it provides rich information about the environment at a relatively low cost and can be used indoors and outdoors. One important issue for vision sensing is the field of view. In many animals, the visual field of view is wide, rendering, in a single image, a global representation of a significant part of the environment (Milford 2008). By analogy, visual sensors, more generally digital cameras, have valuable advantages to tackle the localization and navigation problem. Indeed, in addition to being relatively inexpensive, versatile, lightweight and having low energy consumption, visual sensors bring a lot of information in only one acquisition within indoor or outdoor environments. High levels of capability have been reached regarding the extraction of visual features, the estimation of geometric measurements and even the gathering of semantic knowledge from acquired visual data. Progress made over the past decades to extract this high-level understanding from digital images or videos has shown that visual sensors could be a robust alternative or substantial asset to aforementioned sensors. Thus, the trend toward the exploitation of cameras in localization frameworks has significantly increased. More precisely, research, development and innovation in the field of computer vision have become mature enough to allow the use of cameras as the main and unique sensor, that is, where no other sensory data are required, as a localization solution. A camera can also be used as a secondary sensor in order to provide assistance when the main sensors fail or are unusable. Roughly speaking, a visual localization process generally requires at least three main steps. First, suitable visual data that relevantly describes the surrounding scene have to be extracted from the acquired image. Depending on the prior knowledge about the environment, these extracted visual data have to be matched with either a map that represents the environment, or visual elements that have been previously seen by the camera. Finally, the camera pose within the scene, in other words its localization, is estimated based on the paired data using solid geometry concepts, projective geometry properties and more recently, machine learning methods. Thus, to perform a localization task using vision, it is mandatory to describe the information contained in the acquired images in order to be able to compare them distinctly afterward. These image descriptions, called visual features, must therefore contain as much information as possible, be as discriminant as possible with respect to the described image, and be fast to calculate and compare. The visual features extracted or computed from the acquired images can be classified into two groups: local features and global features. On the one hand, local visual features can be seen as an abstraction of the image information, highlighting parts of the image that are outstanding. These salient image regions are commonly point features (also known as keypoints or interest points), but may also be shapes such as blobs, edges, regions or lines. Once local visual features have been detected (i.e. the extraction stage), each of them are subsequently described in logically different ways by assigning a distinctive identity that enables their effective discrimination (i.e. the description stage). On the other hand, global features describe the visual information in a holistic manner, considering the image as a whole to produce a high dimensional signature. The efficiency of a visual localization method is generally determined according to three criteria: the accuracy of the resulting pose estimation, the memory required to store data and the computational cost. The performance is evaluated on the accuracy of its pose estimation at time t with respect to a ground truth, when this information is available. The main challenges are caused by environment changes in appearance (e.g. lighting and occlusions) and the displacement of dynamic objects (i.e. temporal and spatial variances). Indeed, since the local or global features have to be matched with previously collected features to solve the localization problem, the detection and description stages should ideally provide highly distinctive visual features, robust and invariant to image transformation such as rotation, scale, affine transformations, illumination, noise and so on. In addition, detection, description and comparison must require low computation time and the smallest computer storage capacity.

Localization and navigation benefit from wide fields of view and, in particular, from full omnidirectional vision (Zheng and Tsuji 1992; Yagi et al. 1994, 1995). Usually, omnidirectional images are images with horizontal/vertical fields of view of no less than $180°$ (continuous or not) and vertical/horizontal fields of view of no less than $40°$ (continuous or not). A restricted field of view can be a limitation in environments with low texture, sparse features and/or occlusions. In this case, omnidirectional imaging allows for robustness. Larger portions of the environment can be reconstructed, and robustness against degenerate motions (e.g. rotation only) can be increased (Caruso et al. 2015). Egomotion can be more stable and robust when estimated using these types of images, since ambiguities between translation and rotation induced flow can be reduced (Baker et al.

2001). Free-space detection and mapping are also more robust, stable and faster with omnidirectional imaging (Lukierski et al. 2015). Omnidirectional vision has multiple advantages for autonomous vehicles, since a 360° field of view maximizes information about the environment. A wide field of view significantly decreases the probability of perceptual aliasing (i.e. the possibility that two different and distinct places yield the same sensory reading). In addition, in tracking in a dynamic environment, the continuous view of the area facilitates following the objects of interest. Omnidirectional images can be acquired by a variety of methods and systems. Essentially omnidirectional images can be obtained using two types of systems: catadioptric systems (combining lenses and mirrors) and dioptric systems (using only lenses). In addition, both classes of systems can still be divided into two categories: central or non-central systems. Central systems are those whose projection models have a single center of projection, that is, where all projecting rays intersect at a single point; they have a single viewpoint. These two classes of systems can be implemented based on the use of a single camera or multiple cameras. Most commonly used catadioptric configurations with a single camera use a convex conical, spherical, parabolic or hyperbolic mirror. Baker and Nayar (1998) derived the configurations and shapes of mirrors that ensure a single center of projection. Central systems have the advantage that results derived for perspective cameras can be directly applied. Configurations based on multiple cameras require synchronization of the cameras to guarantee that the images from different cameras are acquired simultaneously. Configurations made up of multiple cameras with overlapping fields of view are usually called polydioptric. Mirrors can be used to minimize the number of cameras. Non-central systems are more flexible and easily built since they do not require specific configurations to ensure central projection. For example, mirrors with special profiles, ensuring specific image and projection properties, can be used. Also, systems using only lenses, such as fisheye lenses, are non-central.

6.2. Modeling image formation of omnidirectional cameras

Modeling the image formation of omnidirectional cameras is the first step to any application of such devices. It consists of finding the mapping between the pixel coordinates and the coordinates of the 3D projection ray for every pixel in the image. There are generic approaches for modeling camera image formation (see Grossberg and Nayar (2001); Sturm and Ramalingam (2004); Miraldo et al. (2011)). Specific models will intuitively provide easier relationships between image coordinates and 3D projecting rays. The underlying geometry of omnidirectional cameras depends on the geometry of the image formation device. The more basic approach for getting a wide field of view of the environment is to have a multi-camera system. Although cameras are becoming cheaper, leading to relatively low cost for multi-perspective cameras, the use of multiple camera systems has other drawbacks, such as the synchronization of image acquisition and the need for a well-calibrated rigid body device connecting all the devices. Several researchers are focused on developing other kinds of omnidirectional sensors. These unconventional cameras are obtained by combining cameras with mirrors and using special lenses. This section covers the most significant works on modeling omnidirectional cameras. In section 6.2.1, we start with systems with a single effective viewpoint. Section 6.2.2 addresses the case of models for general omnidirectional image formation. In section 6.2.3, we describe the cases of purpose-designed mirrors and camera setups. Section 6.2.4 describes one of the most used type of omnidirectional systems, namely a camera with fisheye lenses.

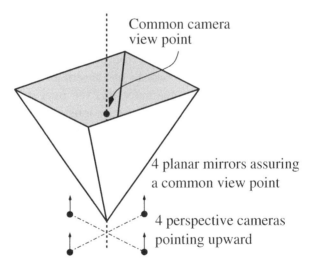

Figure 6.1. *Representation of the first central omnidirectional catadioptric camera system (remastered from Nalwa (1996))*

6.2.1. *Central systems*

Many authors have tried to create omnidirectional cameras while keeping some of the perspective camera constraints. This modeling implies that we can define a one-to-one mapping from the distorted to a perspective image (also known as distortion (Swaninathan et al. 2003)), maintaining all the remaining properties of the perspective camera. The first work presenting a true central omnidirectional catadioptric system was presented in Nalwa (1996). The technical report proposes the use of a pyramid of four planar mirrors and four cameras (a graphical representation of this system is shown in Figure 6.1), with one camera per mirror. Each combination of mirror and camera generates a new view of the environment (the perspective camera viewpoint is changed), which is obtained through the Snell reflection law. By placing the cameras and mirrors in some specific configurations, it is theoretically possible to get a single effective viewpoint, that is, all the reflected rays from the four pairs of perspective cameras and mirrors have a unique viewpoint.

In Baker and Nayar (1998, 1999), the authors derive the central constraints for catadioptric cameras using quadric mirrors, that is, by an implicit equation $\Omega(x, y, z) = 0$, with

$$\Omega(x, y, z) = x^2 + y^2 + Az^2 + Bz - C \qquad [6.1]$$

where $[x\, y\, z] \in \mathbb{R}^3$ represents a 3D point on the mirror (see Figure 6.2). The authors show that by having the camera aligned with the mirror's axis of symmetry and at some specific distance, we can meet a single view-point constraint depending on the mirror's shape. Thus, theoretically, we can have central systems that are conical, spherical, ellipsoidal, hyperboloidal and paraboloidal. More specifically, from [6.1], the authors prove that the following configurations represent central camera systems:

Conical: $A = -\dfrac{2}{k-2}$, $B = 0$, $C = 0$, and $c_3 = 0$; $\qquad [6.2]$

Spherical: $A = 1$, $B = 0$, $C = -\dfrac{k}{2}$, and $c_3 = 0$; $\qquad [6.3]$

Ellipsoidal: $A = \dfrac{2k}{c_3^2 + 2k}$, $B = -\dfrac{2c_3 k}{c_3^2 + 2k}$, $C = -\dfrac{c_3^2 k}{2c_3^2 + 4k} + \dfrac{k}{2}$;

$$[6.4]$$

Hyperboloidal: $A = -\dfrac{2}{k-2},\ B = \dfrac{2c_3}{k-2},\ C = \dfrac{c_3^2 k}{k-2} - \dfrac{c_3^2}{2k};$ [6.5]

Paraboloidal: $A = 0,\ B = \dfrac{2c_3}{k},\ C = \left(\dfrac{c_3}{k}\right)^2$ with $c_3, k \to \infty,$ [6.6]

where $c_3 \in \mathbb{R}$ is the position of the camera in the mirror axis and $k \in \mathbb{R}$ is a mirror parameter. However, since the center of the camera in the conical mirror case must be located at the apex of the cone ($c_3 = 0$), this system cannot be fully obtained in practice. In addition, since in the spherical system all the projection and reflected rays are parallel to each other, there is little room for applications.

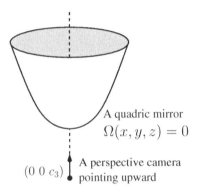

A quadric mirror
$\Omega(x, y, z) = 0$

$(0\ 0\ c_3)$ A perspective camera pointing upward

Figure 6.2. *Representation of an omnidirectional catadioptric camera system with a quadric mirror*

In Nayar (1997), the author presents a new camera system with a hemispherical field of view. Two central catadioptric cameras of the type (Baker and Nayar 1998, 1999) are placed back to back, without breaking the single viewpoint constraint. The author gets at a truly omnidirectional sensor, that is, a 360° field of view. Geyer and Daniilidis (2000) propose a new 2-projection model for the central catadioptric image formation, denoted as $s(.)$ and $p_{l,m}(.)$. The authors assume a unit sphere at the origin, with the z-axis aligned with the perspective camera position, and project the point in the world to the surface of the sphere

$$s(x, y, z) = \left(\pm\frac{x}{r}, \pm\frac{y}{r}, \pm\frac{z}{r}\right),$$ [6.7]

where $r = \sqrt{x^2 + y^2 + z^2}$. Now, considering points $[x\ y\ z]$ on the sphere, the second map is of a perspective projection type

$$p_{l,m}\left(\frac{x(l+m)}{lr-z}, \frac{y(l+m)}{lr-z}, -m\right),$$ [6.8]

where l and m are two-parameter defining the perspective projection mapping (see Figure 6.3). The values for l and m depend on the shape of the mirror. For more details, see Geyer and Daniilidis (2000). This is one of the most used exact models for the image formation of central catadioptric cameras. It consists of a two-projection method. First, we project the 3D points into a sphere of radius one. Then, points on the sphere are projected into the image plane using a perspective projection.

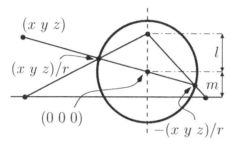

Figure 6.3. *xz-Cut representation of the two projections model for general central catadioptric cameras (from Geyer and Daniilidis (2000))*

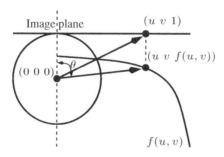

Figure 6.4. *xz-cut representation of general distortion model*

Although not exact, other approaches try to model distortions. Most methods follow this parameterization (naive representation):

$$(u\ v\ 1) \mapsto (u\ v\ f(u,v))$$ [6.9]

where $f(u,v)$ models the distortion, and u and v represent the coordinates of the projection point without distortion. This tells us that the perspective projection ray $(u\ v\ 1)$ is mapped into the distorted projection ray $(u\ v\ f(u,v))$. Fitzgibbon (2001) focuses on small lens distortions, that is, small fields of view. The authors present the division model, with a single distortion parameter model, more specifically, $f(u,v) = 1 + \lambda r^2$, where $r = \sqrt{u^2 + v^2}$. The paper also shows the model's applicability to two important applications, estimation of the fundamental matrix and distortion parameter and also homography estimation. Because of its simplicity and high applicability, this is one of the most used models for modeling central cameras with medium distortion levels. In Micusik and Pajdla (2003), the authors extend the division model to work with a larger field-of-view camera. Instead of a single parameter, the proposed model has more parameters and a nonlinear distortion function, and it is represented as follows: $f(u,v) = \frac{r}{\tan(\theta)}$, where $\theta = \frac{ar}{1+br^2}$, a and b are two additional parameters. The nonlinear nature of the distortion function makes it difficult for many applications. As a result, the authors present a linear approximation of that distortion function using the Taylor series. Another well-known model for omnidirectional sensors was presented in Scaramuzza et al. (2006). It considers the mapping from [6.9], but in this case representing the distortion function as $f(u,v) = a_0 + a_2 r^2 + \cdots + a_N r^N$, whose coefficients are estimated by solving a four-step least-squares linear minimization problem, followed by a nonlinear refinement based on the maximum likelihood criterion. In addition to the proposed modeling and calibration methods, the authors developed a toolbox for camera calibration, which is available for download.

Another interesting model for omnidirectional cameras was presented in Thirthala and Pollefeys (2012). The authors propose the 1D imaging model, in which a point is not mapped into a point in the image. Instead, it is mapped into a straight line in the image that passes through its origin. The model proved to be very helpful in some Structure-from-Motion (SfM) applications due to its invariance to the projection model of distortion.

6.2.2. *Non-central systems*

As mentioned in the beginning of this section, modeling image formation for any camera system implies getting a one-to-one mapping between every pixel in the image and its respective 3D projection ray. Multi-camera systems are non-central omnidirectional devices as a result of the displacements between cameras, and also due to the fact that it is physically impossible to have a common effective viewpoint. Depending on the camera in which we are analyzing a pixel, the projection ray is given by the perspective projection ray direction $\mathbf{d}_i$ passing through the camera center $\mathbf{c}_i$ (or ith camera effective viewpoint). Many researchers (e.g. Pless (2003); Hongdong et al. (2008); Lee and Faundorfer (2013); Klingner et al. (2013)) have been using the more general camera model (GCM) (Grossberg and Nayar 2001; Sturm and Ramalingam 2004; Miraldo et al. 2011) to model this kind of imaging device.

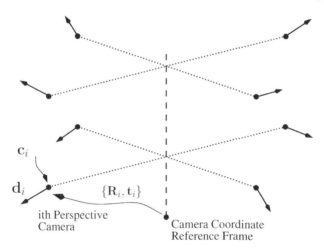

Figure 6.5. *Graphical representation of a multi-camera omnidirectional system with eight perspective cameras*

Concerning the use of catadioptric cameras, and although central catadioptric cameras are theoretically feasible, in practice we cannot ensure that a system is central with a certain accuracy. This is due to the fact that the effective viewpoint of the camera cannot be placed in a specific 3D position and also that there is no real orthographic lens. Indeed, we cannot ensure that the conditions in [6.2]–[6.6] are met. As a result, Swaminathan et al. (2006) define a model for non-central catadioptric cameras. The authors rely on the use of caustics (viewpoint loci) and classify their parameters for the shape of the mirror. Figure 6.6 shows a graphical representation of this model.

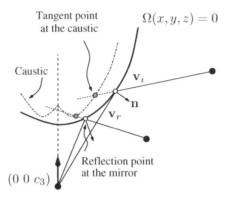

Figure 6.6. *xz-cut of a non-central catadioptric with quadric mirrors. v_i denotes the incident or projection ray, n is the normal vector at the mirror and v_r the reflection ray according to the Snell's reflection law*

While the method in Swaminathan et al. (2006) works for non-central image formation, the nonlinearity involved in the definitions of the caustics makes it difficult to be used in practice. Instead of defining a model for image formation, some authors tried to have specific algebraic mapping functions to make the projection parameterization easier (see Agrawal et al. (2010, 2011)). These works propose the image formation parameterization as a polynomial equation. To define the projection of 3D points, we have to find the roots of the polynomials and then recover the remaining parameters. Agrawal et al. (2010) only consider axial non-central systems. Agrawal et al. (2011) handle any perspective camera position. In Gonçalves (2010), the author claims that the methods proposed in Agrawal et al. (2010, 2011) are slow, mainly because they require the computation of the roots, in which the authors use general solvers, and proposes an alternative iterative method for the same problem. Finally, in Xiang et al. (2013), aimed at relaxing the misalignment between the mirror and the camera, the authors propose a model that is based on the spherical unified camera model from Geyer and Daniilidis (2000), but, instead of considering a projection aligned with the z-axis, it considers a misalignment in the projection of point from the sphere to the image plane. The method proved to be useful for small perspective camera misalignments.

Some authors consider a different type of parametrization to describe non-central catadioptric cameras, that is, they do not consider the underlying geometry of the imaging device explicitly. For example, (Sturm and Ramalingam 2004; Miraldo et al. 2011) use the general imaging models. In Micusik and Pajdla (2004), the authors use an initial estimate for the projection given by a central catadioptric approximation, and the correct projection is optimized using an iterative technique.

6.2.3. *Mirrors with special profiles*

Specific projection models and properties can be obtained by using mirrors with particular profiles. In Chahl and Srinivasan (1997), a family of reflective surfaces is presented that, when imaged by a camera, can capture a global view of the visual environment. The family of surfaces is derived by solving a differential equation expressing the camera viewing angle as a function of the angle of incidence on a reflective surface, as shown in Figure 6.7, where θ is the radial angle, which is the angle of the incoming light rays with respect to the optical axis of the camera and ϕ is the angle of the incoming light rays with respect to the vertical axis of the surface (angle of elevation). A linear relationship between the radial angle and the angle of elevation can be defined as

$$\alpha = \delta\phi/\delta\theta, \qquad [6.10]$$

where α defines the angular magnification in the vertical direction and is constant. To derive the profile of the surfaces, the relationships represented in Figure 6.8 are used. With reference to Figure 6.8, where r is the distance between a point on the surface and the center of projection of the camera and γ is the angle of incidence, we have

$$\tan\gamma = \frac{rd\theta}{dr}. \qquad [6.11]$$

To ensure that a change in the angle of elevation corresponds to a proportional change in the radial angle, the rate of change of γ with respect to θ has to be constant. Therefore,

$$\frac{d}{d\theta}\left[\tan^{-1}\left(r\frac{d\theta}{dr}\right)\right] = \kappa. \qquad [6.12]$$

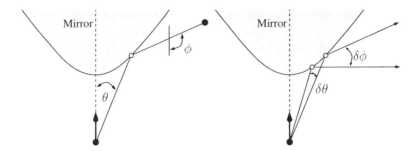

Figure 6.7. *θ denotes the angle of perspective camera projection rays. φ is the angle of incoming light rays, with respect to the vertical axis of the surface*

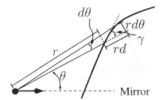

Figure 6.8. *Relationships used in Chahl and Srinivasan (1997) to derive a family of mirror surfaces*

The larger the κ, the greater the rate of change of the angle of incidence with the change of the radial angle. The integration of this equation leads to an expression specifying the profile of the surface. The surfaces preserve a linear relationship between the angle of incidence of light rays onto the surface and the angle of reflection onto the imaging device. However, the gradient of this linear relationship can be varied as desired to produce a larger or smaller field of view. Different values of the gradient lead to different types of wide-angle imaging. This gradient corresponds to an angular magnification, and in this paper several surfaces are derived for different values of angular magnification.

Hicks and Bajcsy (1999) also describe a family of surfaces but, in this case, the images acquired by these surfaces preserve the geometry of planes that are perpendicular to their axes of symmetry. To derive the family of surfaces, this paper considers a configuration similar to the one described in Chahl and Srinivasan (1997) (a pinhole camera pointing up at a curved mirror). The corresponding differential equation is solved by imposing the constraint that the image of planes perpendicular to the axis of symmetry is a scaled version of the 3D plane. The resulting equation is solved numerically. An image obtained with the mirror with the specially designed shape is shown in Figure 6.9, where we see that the image obtained produces a rectified chessboard. In Ollis et al. (1999), an equiangular mirror is described. This means that each pixel in the image spans exactly the same angle. The difference between this configuration and the one described in Chahl and Srinivasan (1997) is that in Chahl and Srinivasan's paper, a small angle approximation is used, which is not the case in Ollis et al. (1999). (Hicks and Bajcsy 2000) extends (Hicks and Bajcsy 1999) by showing that in the case of orthographic projection between the mirror and the camera, the overall transformation and mapping between the plane orthogonal to the symmetry axis and its image approximates a perspective image of the 3D plane.

Figure 6.9. *On the left, the setup considered in Hicks and Bajcsy (1999) is shown. On the right, we have the resulting image obtained by the mirror with specifically designed shape*

A mirror profile to implement a cylindrical projection is derived in Hicks et al. (2001). For this purpose, a specific constraint is applied (defining a linear relationship between the radial distance to the mirror's axis and the height of the point) applied to the differential equation. The camera projection is assumed to be orthographic. A general approach to derive pre-defined projections and, therefore, the corresponding mirrors, is described in Hicks and Perline (2001). The method is based on the concept of distributions, which are generalizations of the concept of differential equations. The idea is that a specific projection that images the world in a specific way corresponds to the determination of the orientation of the tangent planes to the surface. This leads to a pair of partial differential equations, which may or may not have a common solution. This common solution would then determine the mirror surface. If no common solution exists, then an approximation can be obtained using a least-squares-based approximation. This approach was extended and adjusted in Hicks and

Perline (2004). In this paper, the method of vector fields is introduced. The goal is also to realize a specific projection model. The problem is formulated as an estimation of a vector field normal to the mirror surface to be estimated

$$\mathbf{W}(\mathbf{r}) = \frac{\mathbf{q}(\mathbf{r}) - \mathbf{r}}{|\mathbf{q}(r) - \mathbf{r}|} + \frac{G(\mathbf{q}(\mathbf{r})) - \mathbf{r}}{|G(\mathbf{q}(\mathbf{r})) - \mathbf{r}|},$$ [6.13]

where $\mathbf{q}$ is a point in the image plane corresponding to the $\mathbf{r}$, $G(\mathbf{q})$ is a tangent point on the surface and $\mathbf{W}$ is the vector field, as shown in Figure 6.10. This leads to solving partial differential equations that can be solved linearly in some specific cases (in the case where the vector field is a gradient). The paper demonstrates several new sensor designs, including a mirror that yields a panoramic view without any unwarping. An extended version of this paper was published in Hicks (2005). In Gaspar et al. (2002), an approach to design mirrors with a constant vertical resolution, a constant horizontal resolution, or a constant angular resolution was presented (see Figure 6.11). For mirrors with a constant vertical resolution, objects at a (pre-specified) fixed distance from the camera's optical axis will always be the same size in the image, independent of its vertical coordinates. Mirrors with constant horizontal resolution ensure that the ground plane is imaged under a scaled Euclidean transformation. Mirrors with a constant angular resolution ensure similar properties to those of a spherical mirror.

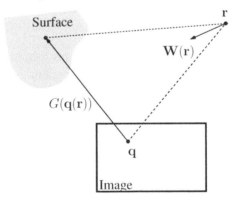

Figure 6.10. *Relationship between the image plane and an object surface and the determination of the vector field **W**(**q**). For a color version of this figure, see www.iste.co.uk/vasseur/omnidirectional.zip*

A family of catadioptric sensors forming panoramic images of uniform resolution is described in Hicks and Coletta (2013). These are rotationally symmetric mirrors, and they differ from catadioptric sensors, which are equi-resolution concerning the solid angle. In this case, the definition of resolution of a catadioptric sensor is the number of pixels per solid angle allocated by the sensor. This is a pointwise notion, that is, the resolution can vary from point to point. Formally, suppose the sensor is viewed as a projection of the sphere at infinity to the image plane. In this case, the resolution is the determinant of the Jacobian matrix of the projection. In previous equi-resolution mirrors (with respect to the solid angle), the images are unwarped to create traditional panoramic strips. These unwarping maps generally have non-constant Jacobian determinants that result in non-uniform distributions of pixels. In Hicks and Coletta (2013), an equi-cylindrical mirror is created.

(a)   (b)

Figure 6.11. *Camera setup proposed in Gaspar et al. (2002) for maintaining constant vertical and horizontal resolution. (a) On the left, we show the catadioptric system for maintaining constant resolution proposed in Gaspar et al. (2002). On the right, the system mounted on a mobile robot is shown. (b) Constant vertical resolution and horizontal resolution images obtained from the system proposed in (a). For a color version of this figure, see www.iste.co.uk/vasseur/omnidirectional.zip*

6.2.4. *Fisheye lenses*

Another class of imaging devices that are commonly used in robotics are cameras with fisheye lenses. Instead of using mirrors for obtaining a wide field of view of the environment, these camera systems use particular kinds of lenses placed in a specific order (dioptric systems). When comparing fisheye lens cameras with catadioptric systems, while the former has the advantage of being a rigid system, their field of view is lower than the latter. In theory, catadioptric systems can have up to 360° of field of view. Concerning the models for these two different imaging devices, Ying and Hu (2004) aim to check whether fisheye camera models can be modeled by central catadioptric. The authors claim that they can and use the model in Geyer and Daniilidis (2000) to represent fisheye imaging devices. Most of the roboticists and computer vision researchers have used constraints from central systems to model these devices. One of the first methods is proposed in Miyamoto (1964). The authors proposed the rectilinear projection mapping function, which has been used in the literature to model distortions with different mapping functions (two-step projection, first onto a sphere and then reprojected into the image plane according to some mapping function). In Kannala and Brandt (2006), the authors propose two models, a radially symmetric one and a model that considers misalignments of the camera components, which makes the distortion not radially symmetric. The authors also derived a backward projection model and a method to obtain the calibration parameters. Several other authors have tried to use central models for representing fisheye cameras. Many different methods using a single effective viewpoint were published (e.g. Hughes et al. (2010); Usenko et al. (2016)).

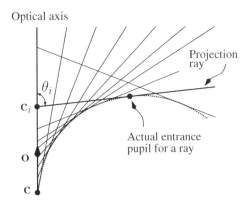

Figure 6.12. *Graphical representation of the varying entering pupil model in Gennery (2006) (non-central camera) for fisheye lens cameras*

Complex lens setups will create a variation on the entrance pupil (camera effective viewpoint). It is assumed that, for conventional cameras, this variation can be neglected, that is, we can assume that the entrance pupil is fixed. While most researchers have assumed that fisheye systems have a fixed entrance pupil (central cameras systems), as shown in Gennery (2006); Fasogbon and Aksu (2019). Having a varying pupil entry (non-central camera system) is more suitable for these kinds of devices. Gennery (2006) presents a model containing a moving entrance pupil, depending on the off-axis angle (i.e. projection line direction). The shifted projecting center, or entrance pupil, is shown in Figure 6.12, where

$$\mathbf{c}_i = \mathbf{c} + s_i \mathbf{o}, \qquad\qquad [6.14]$$

where s_i is a function of the angle θ_i. The modeling focused on the parameterization of this function. The authors use a polynomial representation with correction

$$s_i = \left(\frac{\theta_i}{\sin\theta_i} - 1 \right) \left(\varepsilon_0 + \varepsilon_2 \theta_i^2 + \varepsilon_4 \theta_i^4 + \dots \right), \qquad [6.15]$$

where ε_i are camera calibration parameters. Also using a varying entrance pupil, Fasogbon and Aksu (2019) consider a different approach. Instead of directly parameterizing the $\mathbf{c}_i$ as a function of θ_i, the change in the pupil variation is modeled by moving the z-coordinates of the points such that the transformation from the world (here denoted by $[x'\ y'\ z']$) to the camera coordinate system is obtained as follows:

$$\begin{bmatrix} x \\ y \\ z \end{bmatrix} = \begin{bmatrix} \mathbf{R} & \mathbf{t} \end{bmatrix} \begin{bmatrix} x' \\ y' \\ z' + E(\theta) \\ 1 \end{bmatrix}, \qquad\qquad [6.16]$$

where $E(\theta) = \varepsilon_3 \theta^3 + \varepsilon_5 \theta^5 + \varepsilon_7 \theta^7 + \varepsilon_9 \theta^9$. The ε_i are calibration parameters.

6.3. Localization and navigation

The following section aims to draw up a list of methods, techniques and theories developed over the last 30 years that are based on the properties offered by the omnidirectional vision, in order to face the visual localization and navigation challenges. As it is usually done in reviews and surveys (Paya et al. 2017a), the omnidirectional vision-based localization and navigation approaches presented below are classified according to the type of environment representation. Indeed, in map-based approaches, the robot initially knows a reference map that is generally metric or topological. In contrast, in mapless approaches, the robot operates with respect to observed elements in the environment without the need to create a descriptive map. Therefore, SLAM (simultaneous localization and mapping) is a capability that is central, since it enables the estimation of the robot pose, while at the same time performing the estimation of the environment map.

6.3.1. *Metric localization and mapping*

In this section, approaches based on metric maps are described. A metric map represents the environment as accurately as possible with geometric spatial information, usually called landmarks, beacons or markers, expressed in a reference coordinate system. Different techniques have been proposed to metrically solve the localization problem based on correspondences between the extraction of these landmarks' projection in an acquired omnidirectional image and the relative map. Depending on the number and the nature of its landmarks (e.g. measurements, distances, shapes, sizes and so on), metric maps may have different degrees of details and sizes. A simple metric map can be built by measuring the absolute positions of identical or dissimilar artificial landmarks that can be perceived from a variety of poses in the environment beforehand. The key advantages of using an omnidirectional camera in this context is that many landmarks can be perceived in one single acquisition because of its very wide field of view, moreover the bearing measurements of the landmarks are robust and easily extracted.

Classical geometrical landmark-based localization approaches are based on the principle that two bearing measurements constrain the camera's position to lie on a specific arc of a circle spanned by these landmarks.

One of the very first works describing an omnidirectional vision for localization is by Cao et al. (1985). In this paper, a fisheye lens is used to obtain omnidirectional images. The localization is performed using spherical beacons, with colored lights being used to facilitate segmentation. Three beacons are used, and the camera optical axis is positioned orthogonal to the ground plane. The absolute position and orientation of the robot relative to the world coordinate system would thus be estimated.

A unique solution can be determined by well-known triangulation techniques (Betke and Gurvits 1997) from the perception of third or more landmarks. In Li et al. (2010), a set of four artificial landmarks placed at a known distance from each other in the corners of a rectangular farm field are used to localize an agricultural vehicle equipped with an omnidirectional camera. The shape and colors of the artificial landmarks were chosen to facilitate the detection of their projections in the omnidirectional images with classical image processing techniques. The position of the vehicle is determined from the geometric relation that relies on the four angles formed by the four detected landmarks and their intersection in the omnidirectional image. It has been shown that in a similar environment represented by the same type of map, a solution to the localization problem can be obtained using only the distances and bearings of two detected landmarks (Calabrese and Indiveri 2005). Some works use even simpler metric maps composed of only one landmark. A multicolored three-dimensional shape has been especially designed in Jang et al. (2005) to directly encode its relative distance and orientation in its omnidirectional projection. A coarse estimation of the camera position and orientation can be computed by solving simple trigonometry equations from geometric relations between three points directly extracted from the projection of the proposed landmark. An incremental localization process based on extended Kalman filtering is also introduced to gradually improve the accuracy of the localization as the camera is moving. In Wu and Tsai (2009), a single circular-shaped landmark placed on the ceiling of an indoor environment is used to localize a mobile robot with an embedded omnidirectional camera pointing up. Authors have theoretically and experimentally shown that the projection of a circular-shaped in an omnidirectional image is well approximated by an ellipse following a Taylor series expansion. The localization resolution is analytically obtained from the parameters of the captured elliptical shape, that is, the location of the landmark, including its distance and orientation, relative to the camera. From a practical point of view, this method has been used to localize a helicopter with respect to a helipad preparing for landing (Wu and Tsai 2010). Localization estimated with these artificial landmarks-based approaches is very sensitive to different measurement noises, that is, estimation errors of the angles or distances between the perceived landmarks, inaccuracies during the map measurement itself, misidentified landmarks and so on. These different sources of problems have been studied, and some solutions including hardware and map calibration have been proposed in Loevsky and Shimshoni (2010). Using artificial landmarks offers several advantages such as a simplified detection, identification, tracking and may include direct information regarding their relative localization with respect to the camera. However, this requires preparing and modifying the environment in which the camera operates, which is not always straightforward and is sometimes even impossible.

A second alternative is thus to use natural landmarks that intrinsically represent the environment. Natural landmarks are chosen according to their distinctive characteristics once projected into the omnidirectional image (considering their geometrical and/or photometrical features). For example, since man-made environments generally contain many lines, for example, buildings, windows frames, doors, corridors and so on, they can be used as discriminant visual features. To do so, a 3D representation composed of the main lines of the environment is generally pre-built, then 2D-3D line correspondences are used to solve the localization problem. Natural vertical landmarks have been used in many works since they offer several advantages when used in combination with omnidirectional acquisitions. In a set of papers published in 1990 and 1991 (Yagi and Kawato 1990; Yagi et al. 1991a, 1991b, 1991c; Yagi and Yachida 1991), a research group in Japan describes the development of the COPIS, a catadioptric omnidirectional vision system based on a conic mirror and its application to multiple navigation problems. In Yagi et al. (1991b), the location of the robot is estimated assuming previous knowledge of the environment map. Location is estimated relative to the predefined structures of the environment. The authors use vertical lines in the word, which are projected into straight lines in the image passing through the origin. Indeed, when these setups are vertically oriented, every 3D point with a same azimuth is projected on a same radial straight line in the 2D image and passes through the image center of projection. Unknown objects are assumed to be obstacles, and their positions are estimated using two images and triangulation. In Yagi et al. (1991c), free space is estimated based on the detection and mapping of objects relative to the robot. The approach is based again on the detection of vertical edges in the environment. Essentially, this approach is equivalent to a map estimation since it is based on the detection of objects.

Similarly, another omnidirectional visual system composed of a conic mirror in front of a classical camera, the SYCLOP, has been developed and used in Marhic et al. (1998). The authors proposed to model the 3D environment as an absolute map of reference points that represent the natural vertical landmarks. The authors have demonstrated that the relation between a line that connects two reference points with its projection in the image is actually a 1D projective mapping. The matching between the map and the lines extracted from an acquired image is determined by comparing cross-ratios computed from any combinations of straight lines quadruplets. A Newton–Rapson method is finally used to find a solution to the over-determined system of nonlinear equations. A localization method within a well-structured environment that can be represented as a flat surface composed of straight lines has also been proposed in Marques and Lima (2001). In this work, the mirror of the catadioptric system is specifically shaped to preserve the geometry of the orthogonal ground surface. In this configuration, the camera produces what is generally called a bird's eye view (BEV) of the surrounding environment. The orientation and the position of the camera with respect to the known map is determined using a Hough localization method. Correspondences between the image and the known map are determined based on a Hough transform on the intersections between the detected lines in the captured image and concentric circles around the projection center. The actual localization is addressed by evaluating the probability that the camera is at a certain location given these matches. In Menegatti et al. (2004), a similar environment is represented with two metric maps that depict the surrounding static obstacles and the relevant chromatic transitions of the ground surface. The authors proposed to simulate a $360°$ laser rangefinder in the omnidirectional image. Simulated laser rays are launched around the projection center of the omnidirectional image, then for each ray, the distance between the center and the closest encountered chromatic transition is measured. A slightly adapted Monte–Carlo localization technique is used to compute the probability density of robot position and recursively propagates this probability density using motion and perception information. In addition, this approach has the advantage of being able to easily detect occlusions in dynamic environments. The idea has been extended and validated in general indoor environments with natural color transitions (Menegatti et al. 2006).

In Miraldo and Araújo (2014); Miraldo et al. (2015), the authors assumed a known 3D structure of the environment with 3D lines at the edges. With the known 3D line coordinates, the authors model omnidirectional cameras using the GCM. With pixels defining the line images and their respective 3D projection rays, the authors model planar and general localization as an optimization problem of finding the rotation and translation that aligns the 3D inverse projection rays with the known 3D coordinates of the lines in the world. Some results are shown in Figure 6.13, in which the authors used a spherical catadioptric camera on a mobile robot navigating in the environment with known 3D line coordinates. Vanishing points and lines are geometric concepts that are widely used in localization. While the former corresponds to images of 3D points corresponding to the intersection of parallel lines in the world (a point in the infinite), the latter corresponds to curves whose image points correspond to vanishing line directions. For example, the image of the intersection of every pair of parallel lines in a 3D plane originates a vanishing curve. Since the vanishing points only depend on the direction of the lines, a set of two vanishing points can be used to define the relative orientation from the world to the camera. In Miraldo et al. (2018), the authors model these two geometric concepts for general catadioptric omnidirectional cameras. The authors show some examples of a 3D localization at the top (with an augmented reality example) and the sky segmentation with vanishing lines using non-central catadioptric cameras (see Figure 6.14).

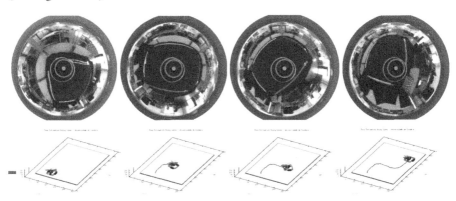

Figure 6.13. *Application of the localization method proposed in Miraldo and Araújo (2014), on a robot equipped with a spherical omnidirectional catadioptric camera. The top row shows the obtained images and the respective fitted line images. The bottom row shows the respective 3D localization and the identification of the environment with the respective 3D lines. For a color version of this figure, see www.iste.co.uk/vasseur/omnidirectional.zip*

In some cases, the environment can be represented by a more descriptive metric map. An approach for localization and navigation based on images acquired by a spherical mirror is described in Hong et al. (1991) (see Figure 6.15). A model of the environment is obtained by having a robot moving along a pre-defined route and acquiring location signatures for specific target locations. These target signatures are then used to locate the robot and compute the robot motion. The target signatures are computed based on specific characteristics of the spherical image, namely by sampling images of horizontal lines and their projective invariance.

Figure 6.14. *Results of the vanishing points and vanishing lines modeling in Miraldo et al. (2018). On the left, we show the modeling of vanishing lines with two catadioptric systems and an example of AR showing that the camera is being localized correctly. On the right, we use the vanishing line modeling to segment sky in an omnidirectional camera. For a color version of this figure, see www.iste.co.uk/vasseur/ omnidirectional.zip*

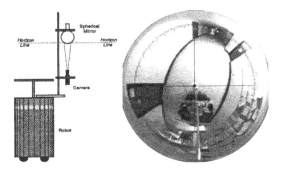

Figure 6.15. *Spherical catadioptric system proposed in Hong et al. (1991), and the respective image of a corridor environment*

In Ramalingam et al. (2010), the authors proposed an outdoor localization method that uses a 3D model of an urban scene composed of untextured meshes. The global localization of the camera is estimated by aligning the skylines extracted from, respectively, omnidirectional real acquisitions facing up and synthetic omnidirectional images rendered within the coarse 3D city map. The actual camera pose is estimated from collinearity constraints given by 2D/3D points and lines correspondences extracted from the matched skylines.

Instead of salient geometric natural landmarks that can be difficult to extract, match and track in the omnidirectional images, persistent and easily observable high-level semantic information of an environment can also be used. The position of static urban objects such as lamp-posts, street-signs, trees, and ground surface boundaries such as manhole covers, pavement edges and lane marking are used in Jayasuriya et al. (2020) to pre-build an outdoor environment. Then, a convolutional neural network (CNN) is used to detect the currently perceived environmental landmarks from an omnidirectional image. The localization is carried out in an extended Kalman filter (EKF) framework using the sparse 2D map and the latter environmental observations.

The manual construction of such maps that contain sufficient geometric or semantic natural landmarks of a scene can be a tedious and complex task. Using external measurement tools can therefore be very valuable. For example, 3D laser scanners have been consistently improved over the last few years and now even include a digital camera that provides color information for each measured 3D point. It is thus now possible to rapidly obtain a dense and precise 3D representation of an environment that contains both geometric and photometric details. A localization method based on these dense colored 3D point clouds as maps from an acquired omnidirectional image has been proposed in Crombez et al. (2015). The camera localization is computed under the visual servoing framework based on photometric features, that is, the visual alignment of the whole virtual and real omnidirectional images, leading to consistent and precise pose estimations. This method has been recently extended to photometric Gaussian mixtures features to dramatically enlarge the convergence domain of the pose estimation (Guerbas et al. 2021).

6.3.2. *Topological localization and mapping*

In this section, approaches based on topological maps are described.

Topological maps represent the environment as a graph, where nodes are specific areas and edges denote topological relationships between each node. They offer an interesting alternative to conventional metric maps because of their simplicity, their rapidity of utilization, their reduced storage requirements, as well as their ease of interpretation for humans. However, a metric map is generally more accurate and therefore preferred for applications that include interactions with the environment (e.g. picking an object, avoiding a person and so on).

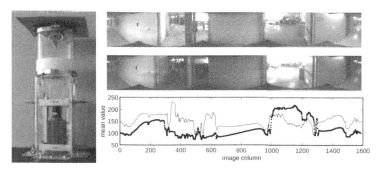

Figure 6.16. *Camera setup used in Winters et al. (2000) and the comparison between a query and references images in the topological map*

Topological navigation using omnidirectional images is described in Winters et al. (2000). A training phase was used to acquire the omnidirectional images (using a spherical mirror) (see Figure 6.16). These images are compressed using principal component analysis. A topological map is built by means of a graph where nodes represent places (identified by images) where actions are undertaken (e.g. turn left). To navigate along the links of the graph, it was assumed that environmental image features could be used for servoing, and the robot position was estimated by determining a reference image that best matched a specific image (as shown in Figure 6.16) within the low dimensional eigenspace

$$d_k = \frac{\sum_{j=1}^{M} (\delta_j^k)^2 \lambda_j}{\sum_{j=1}^{M} \lambda_j},$$ [6.17]

where d_k represents the distance between the current and reference image (all available), δ_j^k is the distance between an image and a specific reference along the jth eigenspace reference and λ_j are the eigenvalues expressing the relative importance of the respective direction in the entire eigenspace. Since an omnidirectional image contains a more complete description of a surrounding scene, a fewer number of views is required to create an extensive representation of a complete environment (Cauchois et al. 2002).

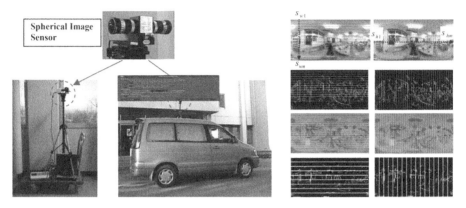

Figure 6.17. *On the left, we show the setup of two fisheye systems in Li and Isago (2007). On the right, we present the pipeline used to obtain the feature space: extracted horizontal and vertical stripes, along the latitude and longitude; the histogram of the orientation of image gradient hue, saturation, and edge points. For a color version of this figure, see www.iste.co.uk/vasseur/omnidirectional.zip*

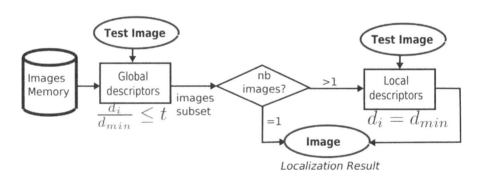

Figure 6.18. *Hierarchical approach for the localization of a robot based on a set of omnidirectional images in Courbon et al. (2008a). For a color version of this figure, see www.iste.co.uk/vasseur/omnidirectional.zip*

Qualitative localization is also exploited in Li and Isago (2007), where an omnidirectional system made up of two fisheye lenses is used, as shown in Figure 6.17. A latitude-longitude image, obtained by Mercator projection, was generated, and a feature space based on horizontal and vertical stripes, extracted along latitude and longitude, was defined. The histogram of image gradient orientation, hue, saturation and edge points make up the feature space (see Figure 6.17.) To perform localization, the robot is guided by humans and memorizes the route scenes by capturing a spherical image sequence. Localization is estimated by comparing the current image with the stored images. Finally, similarity is computed based on a metric applied to the feature space elements. Courbon et al. (2008a) describe an approach for the localization of a robot based on a set of omnidirectional fisheye images (see the pipeline in Figure 6.18). A global descriptor allows a set of candidate images to be selected, after which local descriptors are used to choose the best image. The global descriptor is generated by having a cubic function interpolating the gray level values of each omnidirectional image. The distance between images is computed based on values computed from the interpolated representation. The local descriptor is based on zero normalized cross-correlation between patches around Harris corners. Courbon et al. (2008b) describe a method for the management of visual memory for robot navigation. It is made up of omnidirectional images. The visual memory structure is defined as a multi-graph whose vertices are key images linked by edges, as shown in Figure 6.19. The edges are the visual paths. Images corresponding to two successive nodes of a path contain a set of matched visual features. Localization consists of finding the image of the memory that best fits the camera's current image.

Localization with omnidirectional images and depth data is described in Meilland et al. (2015). The approach is based on three-dimensional omnidirectional visual maps of large-scale unstructured environments. Spherical key-frames contain the light field of all viewing directions from a particular point in 3D space along with its depth map. The depths associated with each pixel are obtained from dense stereo matching. To acquire images, multi-camera systems are used. These configurations allow spherical photometric panoramas with depth information associated with each pixel of the spherical panorama. Each spherical key frame contains the intensities, the depth map and a measure of saliency. A topological graph representation encodes the viewing trajectory. Localization is performed by minimization of an error function that includes the information of the spherical key-frames. Mapping of free space using omnidirectional images is described in Lukierski et al. (2015). For that purpose, a robot acquires a set of closely spaced omnidirectional images. The authors use the unified model proposed in Geyer and Daniilidis (2000) (see section 6.2.1). These images are then unwrapped to a spherically mapped panoramic image. The keypoint-based omnidirectional structure from motion and bundle adjustment is used to register all image frames globally and estimate accurate poses (see Figure 6.20), by obtaining $\mathbf{R}_j$, $\mathbf{t}_j$, and $\mathbf{l}_i$, which minimizes

$$\sum_{i=1}^{n}\sum_{j=1}^{m} b_{ij} \left\| \begin{bmatrix} \frac{1}{\sigma_v} & 0 \\ 0 & \frac{1}{\sigma_v} \end{bmatrix} \mathbf{h}_s \left(\mathbf{R}_j^T \mathbf{l}_i - \mathbf{R}_j^T \mathbf{t}_j \right) - \tilde{\mathbf{u}}_{ij} \right\|_U +$$

$$\sum_{j=1}^{m-1} \left\| \frac{1}{\sigma_d} \left(\|\mathbf{t}_j - \mathbf{t}_{j+1}\|_2 - \|\mathbf{o}_j - \mathbf{o}_{j+1}\|_2 \right) \right\|_C \quad [6.18]$$

where $\{\mathbf{R}_j, \mathbf{t}_j\}$ and $\mathbf{l}_i$ represent the jth camera and the ith landmark pose with respect to the world. $\tilde{\mathbf{u}}_i$ denotes the keypoint detection of $\mathbf{l}_i$. b_{ij} is either 0 or 1, reflecting matched features (i.e. feature i seen on frame j or not). $\mathbf{o}_j$ are the odometry reading, σ are standard deviation parameters, and $\|.\|_H$ and $\|.\|_C$ denote the Huber and Cauchy norms. This problem was solved using the iterative Levenberg–Marquardt algorithm. Then, 3D semi-dense reconstruction is performed and, based on that, a 2D occupancy map is estimated (see Figure 6.20, middle and right). From the semi-dense depth map, the free space area around the reference frame camera pose can then be inferred. Global appearance is used in Paya et al. (2017b) to estimate the position and orientation of a robot in a map. In addition, a global description method based on Radon transform is used. The environment is represented by a topological model built using a set of omnidirectional images and their descriptors. Finally, the localization and orientation are computed based on the distances between the descriptors. For the experiments, the authors used a catadioptric omnidirectional system, represented by the GCM.

Deep learning for place recognition and localization is described in Wang et al. (2018). The approach is used in an application where a robot is only given a few place exemplars, and its current location is unknown and away from these exemplars. The approach allows the robot to retrieve the closest place exemplar and navigate to the nearest place. The approach is based on a CNN that is applied to omnidirectional images. The structure of the network has elements adjusted to the nature of the images. Circular padding is applied to both image and CNN feature spaces to reflect the fact that omnidirectional images have no true image boundary. In addition, a roll branching approach is applied to conquer the rotational invariance in the omnidirectional images.

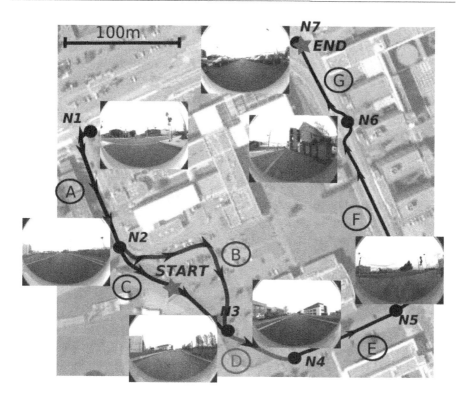

Figure 6.19. *Example of a mapping with the memorized trajectories scheme proposed in Courbon et al. (2008b). For a color version of this figure, see www.iste.co.uk/vasseur/omnidirectional.zip*

Figure 6.20. *On the left, we show the scheme of the problem studied in Lukierski et al. (2015), including a multi-camera pose bundle adjustment. The 3D semi-dense reconstruction is shown in the middle and the respective 2D occupancy map is presented on the right. For a color version of this figure, see www.iste.co.uk/vasseur/omnidirectional.zip*

6.3.3. *Visual odometry*

Visual odometry consists of the estimation of the motion of a vehicle using only vision (Siegwart et al. 2011). Visual odometry is based on structure from motion. From the relationships that model structure from motion, the rotation and translation between consecutive frames are computed, and the vehicle trajectory can be estimated. In many cases, iterations of two-view structure from motion are used. The main goal of visual odometry is the estimation of vehicle trajectory. However, and in many cases, a 3D map of the environment is also estimated, based on simple triangulation of the feature points used to estimate motion.

Omnidirectional images are also advantageous for visual odometry since the extended visual field allows for robustness against occlusions and possible lack of texture in some areas of the visual field. Early approaches that assume the existence of a single effective viewpoint (i.e. the condition presented in Baker and Nayar (1999)) used specialized versions of the essential matrix for perspective cameras $\mathbf{E} \in \mathbb{R}^{3 \times 3}$, such as

$$\mathbf{d}_2^T \underbrace{[\mathbf{t}]_{\times} \mathbf{R}}_{\mathbf{E}} \mathbf{d}_1 = 0, \qquad\qquad [6.19]$$

where $\mathbf{d}_1, \mathbf{d}_2$ are projection rays from the two cameras, or specific models that accounted for the particular projection model of the vision system. After the derivation of the generalized epipolar constraint and the corresponding generalized essential matrix (GEM) (see Pless (2003); Miraldo and Araújo (2015); Miraldo and Cardoso (2020)), many approaches were based on GEM. The generalized epipolar geometry was defined using GEM, $\mathcal{E} \in \mathbb{R}^{6 \times 6}$, such that

$$\begin{bmatrix} \mathbf{d}_2^T & \mathbf{m}_2^T \end{bmatrix} \underbrace{\begin{bmatrix} R[\mathbf{t}]_{\times} & R \\ R & \mathbf{0} \end{bmatrix}}_{\mathcal{E}} \begin{bmatrix} \mathbf{d}_1 \\ \mathbf{m}_1 \end{bmatrix} = \mathbf{d}_2^T R \mathbf{m}_1 + \mathbf{d}_2^T R[\mathbf{t}]_{\times} \mathbf{d}_1 + \mathbf{m}_2^T R \mathbf{d}_1 = 0, \quad [6.20]$$

where $[\mathbf{d}_1 \, \mathbf{m}_1], [\mathbf{d}_2 \, \mathbf{m}_2] \in \mathbb{R}^6$ are the two 3D projection rays in *Plücker* coordinates. In some configurations, the use of omnidirectional images may not be advantageous (Zhang et al. 2016). The use of a large field of view allows for tracking visual landmarks over longer periods, leading, in principle, to an increase in the precision of motion estimates due to the availability of more measurements. In principle, it should also lead to improved robustness because the visual overlap between successive images is larger. However, if the image resolution is the same, the increased field of view will decrease the angular resolution of the image. In Zhang et al. (2016), it is shown that in indoor scenarios, omnidirectional cameras outperform the perspective ones since features are more evenly distributed in space, and also due to the fact that the camera can track features for a longer time. In outdoor environments, however, motion can be estimated more accurately using perspective cameras, essentially because the decrease in angular resolution for larger fields of view is further affected by the higher depth range.

A simple pipeline for accurate and robust motion estimation using computer vision is shown in Figure 6.21. It has a sequence of images (they can be either omnidirectional or any other kind of images) as inputs and consists of the following steps: (i) feature extraction that gets a set of important characteristics for every input image; (ii) obtain the correspondences between each pair of images – more than a pair of images can be used in more complex pipelines to get a co-visibility graph; (iii) remove outliers from the matches and obtain the first guess for the motion estimation, usually RANSAC-based techniques; (iv) using the available inliers and the initial guesses for the poses between the images, estimate the motion; and (v) after getting a set of relative poses of the images (up to a scale factor) and performing a simple triangulation, apply an optimization consisting of minimizing the re-projection error of the 3D points, by considering small variations in the coordinates of the 3D points, the relative position of the cameras and the camera parameters. Below, we describe some alternatives for performing visual odometry using omnidirectional images.

Figure 6.21. *Basic pipeline for visual odometry*

One of the first papers reporting the estimation of visual odometry using an omnidirectional vision system was by Bunschoten and Ben Krose (2003), and it focused on the motion estimation step of the visual odometry pipeline (see Figure 6.21). The vision system used in this work was a single center of projection omnidirectional mirror as derived in Baker and Nayar (1999). A version of the essential matrix in [6.19] is used, constrained to planar motion, to compute rotation and translation between each pair of images

$$\mathbf{E} = \begin{bmatrix} 0 & 0 & \sin\phi \\ 0 & 0 & -\cos\phi \\ \sin(\theta - \phi) & \cos(\theta - \phi) & 0 \end{bmatrix}, \qquad [6.21]$$

where ϕ and θ defines the baseline direction and the rotation around the $z-$axis.

Figure 6.22. *Vehicle with a catadioptric omnidirectional system used in Scaramuzza and Siegwart (2008); Scaramuzza et al. (2009) (from Scaramuzza et al. (2009)). For a color version of this figure, see www.iste.co.uk/vasseur/omnidirectional.zip*

The scale of translation is estimated based on homographies (also constrained to planar motion) between images of the floor. In Corke et al.

(2004), an equi-angular omnidirectional mirror is used. This is a non-central system. Two methods for visual odometry are described and compared: one method based on the optical flow estimation and subsequent integration of the flow and one method based on SfM. The SfM method is based on the specific mirror model used. They conclude that the SfM approach produces a higher precision estimation of vehicle motion. In Scaramuzza and Siegwart (2008), a central omnidirectional system is also used to estimate visual odometry. Figure 6.22 shows the vehicle setup used; the image formation is modeled using a general distortion approach (see Figure 6.4). The pipeline follows what is shown in Figure 6.21, with image features being extracted with methods for perspective cameras. Motion estimation is based on a combination of floor-based homography and appearance-based estimation of rotation. Rotation estimates based on the homography are sensitive to systematic errors, and as a result, the rotation is also estimated with an appearance-based approach. In Tardif et al. (2008), an omnidirectional system based on several perspective cameras (a polydioptric configuration) is used. See the setup and resulting images in Figure 6.23. Pose between consecutive images is estimated by decoupling its rotation. Rotation is estimated using robust computation of the epipolar geometry, [6.19] and RANSAC, and only the camera position is determined using 3D points that were previously estimated by triangulation.

An approach for the trajectory estimation of a vehicle based on a central omnidirectional camera is presented in Scaramuzza et al. (2009). The authors use the setup represented in Figure 6.22. A constrained motion model, based on nonholonomic constraints of wheeled vehicles, enables the parameterization of the motion with only one feature correspondence. The constrained essential matrix is used to estimate the motion. More specifically, from the motion model defined in Figure 6.24, the authors adopt the essential matrix representing planar motions in [6.21] to

$$\mathbf{E} = \rho \begin{bmatrix} 0 & 0 & \sin(\frac{\theta}{2}) \\ 0 & 0 & -\cos(\frac{\theta}{2}) \\ \sin(\frac{\theta}{2}) & \cos(\frac{\theta}{2}) & 0 \end{bmatrix}. \qquad [6.22]$$

Since $\mathbf{E}$ can be estimated up to a scale factor (see [6.19]), and this represents a one degree of freedom estimation, it can be estimated with a single feature correspondence.

Figure 6.23. *Vehicle with an omnidirectional multi-camera system and the resulting images (from Tardif et al. (2008)). For a color version of this figure, see www.iste.co.uk/vasseur/omnidirectional.zip*

Mouragnon et al. (2009) describe an approach where the GCM (Grossberg and Nayar 2001; Sturm and Ramalingam 2004; Miraldo et al. 2011) is used to estimate 6D motion using the generalized epipolar geometry [6.20], but also to perform reconstruction. The algorithm operates in real time, and a local bundle adjustment is used, allowing 3D points and camera poses to be refined simultaneously through the sequence by minimizing

$$\mathbf{g}^i(\mathcal{C}^i, \mathcal{P}^i) = \sum_{C^k \in \{C^{i-N+1}, \dots, C^i\}} \sum_{P_j \in \mathcal{P}^i} \|\epsilon_j^k\|^2, \qquad [6.23]$$

every time a new frame i is obtained. $\mathcal{C}^i$ and $\mathcal{P}^i$ are the GCM parameters and the 3D points chosen for i, with C^i being the camera poses, and P_j the 3D points. A generic re-projection error function is represented by ϵ_j^k. ϵ is defined based on back-projected rays and it is optimized by minimization of the angular error between rays. Visual odometry for the case of a multi-camera system with two perspective cameras with non-overlapping fields of view is addressed in Kazik et al. (n.d.) (see the camera setup in Figure 6.25). Up-to-scale 6D motion estimation is carried out in each of the cameras, and the metric-scale is recovered via a linear approach by enforcing the known static transformation between both sensors. Statistical fusion is used to exploit the redundancy in the motion estimates to obtain an optimal 6D metric result. In Lee and Faundorfer (2013), motion estimation is performed based on the GEM (Pless 2003). The approach is applied to an omnidirectional camera system made up of several perspective cameras (see Figure 6.26). The Ackermann motion model is used to constrain the generalized essential matrix. The resulting epipolar constraint is represented as [6.20], with

$$
\mathcal{E} = \begin{bmatrix}
0 & 0 & \rho\sin(\theta/2) & \cos(\theta/2) & -\sin(\theta/2) & 0 \\
0 & 0 & -\rho\cos(\theta/2) & \sin(\theta/2) & \cos(\theta/2) & 0 \\
\rho\sin(\theta/2) & \rho\cos(\theta/2) & 0 & 0 & 0 & 1 \\
\cos(\theta) & -\sin(\theta) & 0 & 0 & 0 & 0 \\
\sin(\theta/2) & \cos(\theta/2) & 0 & 0 & 0 & 0 \\
0 & 0 & 1 & 0 & 0 & 0
\end{bmatrix}, \qquad [6.24]
$$

where θ and ρ are as shown in Figure 6.24. The approach allows for the estimation a single set of rigid motion parameters and allows for the estimation of metric scale when motion is not purely translational. The combination of omnidirectional images acquired with a fisheye lens and inertial information for visual odometry is described in Ramezani et al. (2017). The setup used is shown in Figure 6.27. The approach is based on a multi-state constraint Kalman filter (MSCKF). Instead of considering visual measurements in the image plane, individual planes are used for each point. These planes are tangent to the unit sphere and normal to the corresponding measurement ray. The relation between the image point and the normalized point on the unit sphere is established based on Scaramuzza's distortion model (see Figure 6.9). In this way, spherical images acquired by the omnidirectional camera are combined with inertial measurements within the filtering approach. The results show that omnidirectional visual-inertial odometry, as a result of the wider field of view, allows incorporating more visual features from the surrounding environment, which improves the accuracy and robustness of the motion estimation. An approach for real-time direct monocular visual odometry for omnidirectional cameras is described in Matsuki et al. (2018). The unified omnidirectional model from Geyer and Daniilidis (2000) is used as a projection function (see section 6.2.1), which is applied to fisheye cameras. In addition, the joint optimization of camera poses, point depths and affine brightness parameters for photometric consistency allows for robustness with fisheye images. The quantitative experimental results on two public benchmark datasets demonstrated that the approach yielded better performance than other competing approaches. A system made up of four cameras with fisheye lenses, with overlapping fields of view between neighboring cameras, is described in Seok and Lim (2019). The four cameras are located at the vertices of a quadrangular rig (see Figure 6.29). A specific projection model is introduced (see Figure 6.30) since, in the large overlapping areas between cameras, direct feature matching is difficult as a result of the large distortion. The projection model for each camera includes a cylinder centered at the center of the camera. The projection model has two projection planes on the left and right sides, and the cylinder at the center connects the two planes smoothly. The experimental results show that such a hybrid projection model allows inter- and intra-camera feature matching. Motion is estimated based on multi-view P3P RANSAC. Recently, machine learning-based approaches for visual odometry (e.g. Wang et al. (2018)) started to be proposed and developed. However, relevant omnidirectional image-based approaches have yet to be proposed and evaluated.

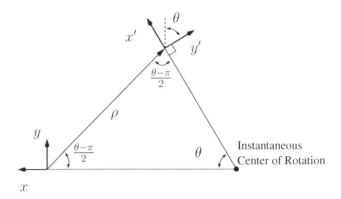

Figure 6.24. *Representation of the motion model used in Scaramuzza et al. (2009)*

Figure 6.25. *Multi-camera system used in Kazik et al. (n.d.). For a color version of this figure, see www.iste.co.uk/vasseur/omnidirectional.zip*

Figure 6.26. *Vehicle with a multi-camera system used in Lee and Faundorfer (2013). Cameras at the front, rear and side cameras in the mirrors. For a color version of this figure, see www.iste.co.uk/vasseur/omnidirectional.zip*

6.3.4. *SLAM*

SLAM stands for "simultaneous localization and mapping". Mapping implies the estimation of the 3D environment, which generally requires the position and path of the robot. Localization requires the estimation of the robot position (and orientation) in a known map. SLAM aims to estimate and recover both the map and the vehicle trajectory at the same time. In general,

SLAM is implemented using data such as images, RGB-D images, point clouds and proprioceptive sensors (e.g. inertial data). SLAM can also benefit from omnidirectional images, since a significant portion of the visual field is represented in each image. In addition, omnidirectional images also facilitate the requirement of having partially overlapped fields of view. SLAM allows for a higher degree of autonomy since it permits the vehicle to operate in situations where map information or vehicle position are not initially known and environment modification is not possible. This ability is essential for fully autonomous behavior and extremely important for a wide range of applications.

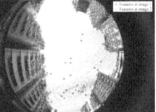

Figure 6.27. *Sensor setup, example of an omnidirectional image with extracted features and the result obtained with the method in Ramezani et al. (2017). For a color version of this figure, see www.iste.co.uk/vasseur/omnidirectional.zip*

One of the very first papers to propose a SLAM-like approach with omnidirectional images was by Strelow et al. (2001). The authors define shape-from-motion, mentioning that "algorithms for shape-from-motion simultaneously estimate camera motion and scene structure". This is what is now known as SfM. This paper describes both batch and online SfM algorithms for omnidirectional cameras and a calibration technique. They use the mirror described in Ollis et al. (1999).

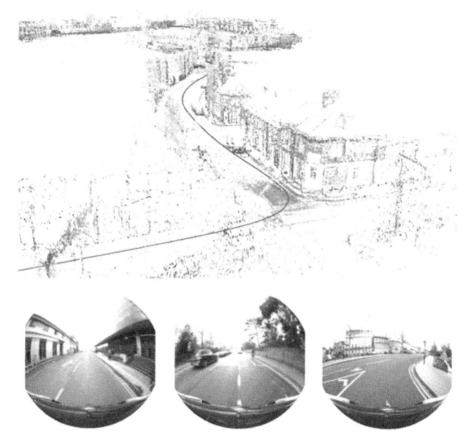

Figure 6.28. *Examples of fisheye images and the resulting trajectory for the method in Matsuki et al. (2018). For a color version of this figure, see www.iste.co.uk/vasseur/omnidirectional.zip*

The system generates the so-called equiangular images. In the batch algorithm, they use a set of images and minimize

$$\chi^2 = \sum_{i,j} \|x_{i,j} - \Pi(\mathbf{R}_j X_i + \mathbf{t}_j)\|^2, \qquad\qquad [6.25]$$

where $\Pi(.)$ describes the 3D projection, X_i is the ith point in the world, and $x_{i,j}$ represents the projection of the point X_i into the image with camera position given by $\{\mathbf{R}_j, \mathbf{t}_j\}$, representing the error between observed feature projections and predicted feature projections with respect to the camera

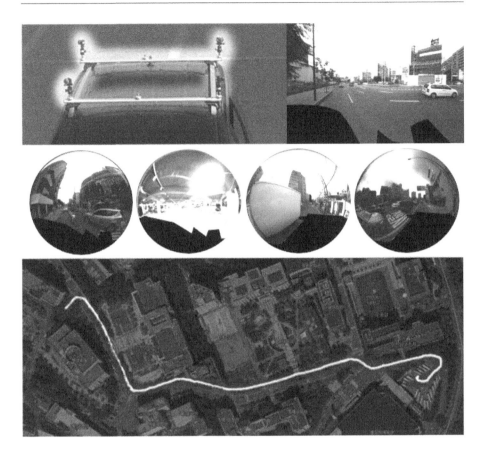

Figure 6.29. *The top picture shows the setup of four fisheye cameras used in Seok and Lim (2019). The middle picture shows four images obtained with the cameras. The bottom picture shows the trajectory obtained with the proposed method. For a color version of this figure, see www.iste.co.uk/vasseur/ omnidirectional.zip*

position, orientation and feature positions. As a result, they obtain the rotation matrix and the translation vector for the camera in the world coordinate system, $\{\mathbf{R}_j, \mathbf{t}_j\}$, as well as the 3D feature position in the world coordinate system. For the so-called "online shape-from-motion", an EKF is used, with the measurement equation being the projection equation, which is nonlinear in the estimated parameters. Another paper proposing SLAM with omnidirectional cameras was by Rituerto et al. (2010a). In this paper, the

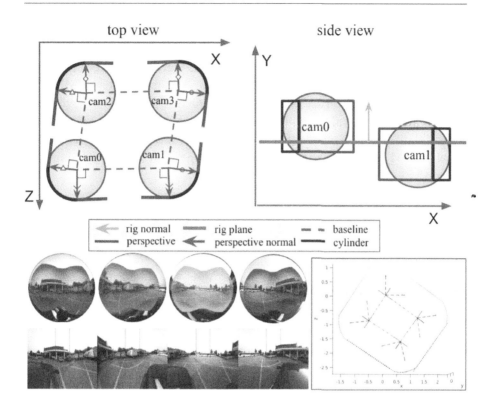

Figure 6.30. *In the top, we show the hybrid projection model proposed in Seok and Lim (2019). The rig plane is the best fitting plane given the relative position of cameras. For a color version of this figure, see www.iste.co.uk/vasseur/omnidirectional.zip*

authors use the spherical camera model by Geyer and Daniilidis (2000) (see section 6.2.1) with an EKF, with system state

$$\mathbf{x}_k = \left(\underbrace{\mathbf{r},\ \mathbf{q},\ \mathbf{v},\ \boldsymbol{\omega}}_{\text{Camera State}},\ \underbrace{x_i,\ y_i,\ z_i,\ \theta_i,\ \phi_i,\ \rho_i}_{\text{Point}}\right)^T, \qquad\qquad [6.26]$$

by means of the Jacobian of the camera model and the Jacobian of the inverse projection. The same authors perform a comparison between omnidirectional and perspective monocular visual SLAM in Rituerto et al. (2010b). Based on tests performed in five trajectories, they conclude that the results obtained with the omnidirectional camera are superior to those obtained with the perspective camera. In Heng et al. (2014, 2015), a multi-camera-based visual

SLAM is described. The approach is based on the generalized camera model. It is a keyframe-based implementation, and a three-point algorithm is used to estimate relative motion with a metric scale. The algorithm makes use of the relative rotation measurement from the inertial measurement unit (IMU) via short-term integration of gyroscopic measurements. A graph of keyframes and constraints obtained from both visual odometry and loop closures is built incrementally. Results show that the approach allows real-time on-board visual SLAM with loop closures on a micro-aerial vehicle with multiple cameras.

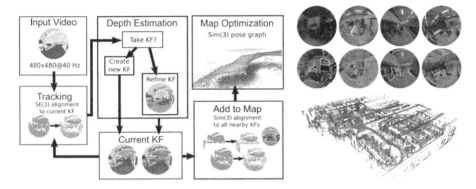

Figure 6.31. *On the left, we show the simultaneous localization and mapping pipeline using omnidirectional images presented in Engel et al. (2014). On the right, we show some experimental results of the application of the pipeline to a real large-scale navigation problem. For a color version of this figure, see www.iste.co.uk/vasseur/omnidirectional.zip*

In Caruso et al. (2015), a method for monocular SLAM for omnidirectional cameras and wide field-of-view fisheye cameras is described. The approach is based on (large-scale direct) LSD-SLAM (Engel et al. 2014) and uses the unified model for central omnidirectional cameras presented in section 6.2.1. The method includes the formulation of direct image alignment for the unified omnidirectional model and a fast approach to incremental stereo directly on distorted images (see Figure 6.31, left). The authors conclude that results obtained with the omnidirectional camera outperform the pinhole, showing that the algorithm benefits from additional information in the image due to an increased field of view, since a wider field of view increases the period during which 3D points are visible. The right side of Figure 6.31 shows some results. Gamallo et al. (2015) address the problem of severe occlusions in omnidirectional visual SLAM. To handle occlusions, a hierarchical data association method is proposed. The algorithm uses visual landmarks and a two-level hierarchical data association, which allows the handling of the occlusion of the landmarks. In Munguía et al. (2017), a filter-based and feature-based SLAM for omnidirectional images is described. This approach uses the unified model for central omnidirectional cameras and requires the depth of the features to be estimated. It is based on an EKF.

Some works have proposed deep learning techniques for obtaining the map of the environment with omnidirectional images. In Tateno et al. (2018), the authors propose a distortion-aware convolution operator. The proposed operator follows a two-step strategy: (1) features are sampled by applying a grid $\mathcal{R}$ to the feature map f_l; (2) the features are weighted using w, such that

$$f_{l+1}(p) = \sum_{r \in \mathcal{R}} = w(r) \cdot f_l(p + \delta(p, r)), \qquad [6.27]$$

where l denotes the network layer and $\delta(p, r)$ is the function representing the distortion-aware convolution grid. The proposed operator can be used in any CNN architecture. In the paper, the authors modify their previous work for perspective cameras (Laina et al. 2016), the fully convolutional residual network, which is applied to SLAM, panoramic semantic segmentation and panoramic style transfer. There are other works applying CNNs to omnidirectional images for various visual tasks. For example, in Lee et al. (2019), the authors consider divisions of a sphere by arcs into bounded regions, using a spherical polyhedron (see the spherical polyhedron in Figure 6.32 for different resolutions). The authors start by projecting the images from the omnidirectional cameras into the spherical polyhedra. Then, the authors adapt the conventional CNNs to SpherePHD; they design special convolution and pooling kernels.

The work in Won et al. (2019) (extended in Won et al. (2020a)) focuses on mapping (i.e. depth estimation) using a wide-baseline multi-view stereo omnidirectional camera setup. The proposed network (see Figure 6.33) takes a set of omnidirectional images as input and outputs the omnidirectional depth estimation. The architecture is divided into three stages: (i) unary feature extraction, (ii) spherical sweeping and (iii) cost volume computation. The first step aims to extract features (using 3D CNNs) directly from the input images, making the network learn the distortion in the original image.

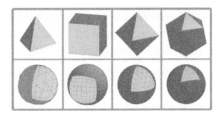

Figure 6.32. *Different resolutions of the spherical polyhedron used to project the images in Lee et al. (2019). For a color version of this figure, see www.iste.co.uk/vasseur/omnidirectional.zip*

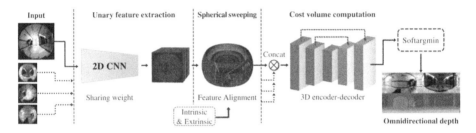

Figure 6.33. *The proposed three-stage DNN (unary feature extraction, spherical sweeping and cost volume computation) for depth estimation using sets of omnidirectional images (see Won et al. (2019)). For a color version of this figure, see www.iste.co.uk/vasseur/omnidirectional.zip*

These generated unary feature maps are then projected onto a predefined sphere using the spherical sweeping module. Finally, the features in the sphere are merged and used as the input of the cost volume computation subnetwork. As for the loss function, the authors used the absolute error (supervised learning):

$$L(\theta, \phi) = \frac{1}{\sum_i M_i(\theta, \phi)} |\hat{n}(\theta, \phi) - \text{round}(n^*(\theta, \phi))|, \qquad [6.28]$$

where $\hat{n}$ and n^* are the estimated and ground-truth inverse of the depth, and θ and ϕ are parameters representing the projection of points into the spherical coordinates. A few works were proposed for depth estimation with omnidirectional images. For example, in Zioulis et al. (2018), the authors start by addressing some issues in the generation of omnidirectional datasets by re-using large-scale 3D datasets and changing them to get a 360° field of view via image rendering. Then, the authors change two encoder-decoder fully convolutional networks to propose end-to-end methods for omnidirectional depth estimation. Similarly to adaptations of perspective datasets, a depth estimation method using deep neural networks (DNNs) is performed in de La Garanderie et al. (2018). In this case, the authors unwarp the images so that they are equirectangular and focus on the issues arising from these resulting images' left and right extremities. To handle the possibility of having an object in the image extremities, the authors propose padding the left and right side of the image with pixels from the right and left side, respectively, as if the image was tiled, which can be seen as a 360 ring.

Figure 6.34. *Fernandez-Labrador et al. (2020) propose a method for mapping the 3D layout of a room by detecting its corners in panoramic images. For a color version of this figure, see www.iste.co.uk/vasseur/omnidirectional.zip*

Some authors try to get room layout mapping from single images. For example, Fernandez-Labrador et al. (2020) propose an end-to-end approach that obtains the layout by detecting corners in panoramic images (see Figure 6.34). As done by some methods in the previous paragraph (e.g. Tateno et al. (2018)), the proposed method adopts the standard convolutions in the base ResNet-50 network to equirectangular convolutions, defined in the sphere instead of the planar image domain, jointly predicting edge and corner maps. In Sun et al. (2019), the authors propose the HorizonNet, a fast network for predicting the 1D Manhattan room layouts. The proposed network is divided into two stages: a feature extractor, in which the authors adopt the ResNet-50 architecture, capturing both low-level and high-level features, and a recurrent neural network for learning patterns and long-term dependencies from sequential data. Many other works were published on this topic. A comparison study on room layout estimation is presented in Zou et al. (2021).

Depth estimation using DNNs was also applied in full SLAM pipelines. For example, the method proposed in Won et al. (2019) is used in a full SLAM pipeline in Won et al. (2020b), as seen in Figure 6.35. The proposed pipeline takes a set of images as input and estimates the omnidirectional depths (Won et al. 2019). Depth is then integrated into localization modules using an adaptation from ROVO (work proposed in Seok and Lim (2019)). Instead of computing the 3D points by triangulation, the authors use the depth estimated from Won et al. (2019), which provides a better, more efficient and robust solution. The loop closing module follows the conventional steps. The authors first train a vocabulary tree to detect loop closures. By running some queries, the authors ensure loop candidates are checked for consistency using a multi-view P3P RANSAC. After detecting the loop closures, the method runs a pose-graph optimization to correct the trajectory.

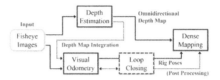

Figure 6.35. *SLAM pipeline proposed in Won et al. (2020b) using the multiple omnidirectional image depth estimation in Won et al. (2019). For a color version of this figure, see www.iste.co.uk/vasseur/omnidirectional.zip*

6.3.5. *Multi-robot formation*

Robot formation aims to ensure that the relative position of a set of agents is held constant while moving in the environment. Usually, we have an agent as a leader. While this one is performing some action, the other agents (followers) will have to move to ensure the formation. We have to make sure that followers are "seeing" the leader when just using onboard sensing. If we use computer vision to localize the agents, having a conventional perspective camera constitutes a hard limitation for the navigation problem, making the multi-agent formation problem significantly harder. Therefore, when using computer vision for the relative agent position sensing, omnidirectional cameras has been used as sensors in some previous studies (Das et al. 2002; Vidal et al. 2003, 2004; Mariottini et al. 2009, 2007).

A framework for formation control of a multi-mobile robot system is presented in Das et al. (2002). The paper introduces a switching method of simple decentralized controllers that allows for changes in formation and obstacles, letting the robots choose the most appropriate controllers depending on the environment (see the controllers used in Figure 6.36).

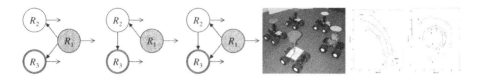

Figure 6.36. *Three different control schemes are considered in the switching method proposed in Das et al. (2002). In the middle, we show the robot setup (each robot is equipped with an omnidirectional catadioptric camera). Some results are shown on the right*

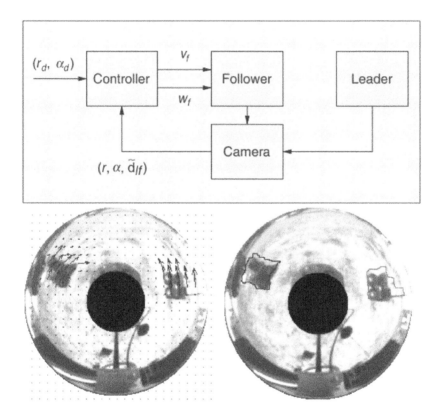

Figure 6.37. *At the top, we show the control scheme used in Vidal et al. (2004). At the bottom, we show an example of optical flow and the respective motion segmentation. For a color version of this figure, see www.iste.co.uk/vasseur/omnidirectional.zip*

One of the claims is the use of a single onboard sensor per robot, relying on the use of central catadioptric cameras (Baker and Nayar 1998), as shown in Figure 6.36. The authors use omnidirectional sensors and a derived obstacle detector, collision detector, decentralized state observer and centralized state observer to allow each robot to estimate the relative position of its teammates up to a scale factor (a direction gives the position of all the other robots). Mariottini et al. (2009) also consider a leader/follower formation problem of only two mobile robots. The authors analyze a generic nonlinear system and define the general observability conditions based on the extended output Jacobian. These results derive the conditions for the leader-follower system to be locally weakly observable and not locally weakly observable, which is afterward used to design an EKF for the leader/follower to control the formation. In Mariottini et al. (2007), the authors extend the previous work by considering that the cameras are uncalibrated, using unscented Kalman filters, and estimating the followers poses using a Kalman filter.

Figure 6.38. *At the top, we show the multi-robot system used in Gava et al. (2007) for formation (with a single catadioptric camera) and the respective detection and tracking module result. For a color version of this figure, see www.iste.co.uk/vasseur/omnidirectional.zip*

Vidal et al. (2003, 2004) aim to perform a decentralized formation control of nonholonomic mobile robots, as shown at the top of Figure 6.37. The authors obtain omnidirectional images with catadioptric camera systems. The control input is modeled in the image by developing an image-based visual servoing scheme. First, the authors use optical flow to estimate the other agents' velocities in the catadioptric omnidirectional image plane (see the bottom of Figure 6.37), which is given by

$$\begin{bmatrix} \dot{u} \\ \dot{v} \end{bmatrix} = \begin{bmatrix} -v \\ u \end{bmatrix} \Omega_z + \frac{1}{\lambda} \begin{bmatrix} 1 - \rho u^2 & -\rho uv \\ -\rho uv & 1 - \rho v^2 \end{bmatrix} \begin{bmatrix} V_x \\ V_y \end{bmatrix},$$ [6.29]

where u and v are the coordinates of points in the image, $\{V_x, V_y\}$ and Ω_z are the robots linear and angular velocities, $\lambda = -z + \xi\sqrt{x^2 + y^2 + z^2}$ and $\rho = \frac{\xi^2}{1+\eta}$ with $\eta = \frac{-1+\xi^2(x^2+y^2)}{1+\xi\sqrt{1+(1-\xi)(x^2+y^2)}}$. ξ is a parameter modeling central catadioptric cameras derived in Geyer and Daniilidis (2000). Using the measured robot's optical flows across multiple frames, the authors describe a robot motion segmentation. The results are shown in Figure 6.37. Then, the authors define central panoramic leader-follower dynamics based on a unicycle motion model. The equations of pixel motion expressed as drift-free control systems are given by

$$\begin{bmatrix} \dot{u} \\ \dot{v} \end{bmatrix} = \begin{bmatrix} \frac{1-\rho u^2}{\lambda} & -v \\ \frac{-\rho uv}{\lambda} & u \end{bmatrix} \begin{bmatrix} v_f \\ \omega_f \end{bmatrix} + \begin{bmatrix} \frac{1-\rho u^2}{\lambda} & \frac{-\rho uv}{\lambda} & -v \\ \frac{-\rho uv}{\lambda} & \frac{1-\rho v^2}{\lambda} & u \end{bmatrix} \begin{bmatrix} F_{lf} \\ \omega_l \end{bmatrix}$$ [6.30]

where $\{v_f, \omega_f\}$ are the control inputs for the follower, $\{F_{lf}, \omega_l\}$ can be seen as an external input that depends on the leader and follower, and on the control actions of the leader. Two visual servoing types of control schemes are proposed, a linear one, by applying feedback linearization, and a nonlinear one to move the current position of the neighboring robots in the image to the desired one.

Instead of having one omnidirectional camera per robot, Gava et al. (2007) propose a centralized formation control in which only the leader robot has a camera. The leader performs robots detection and tracking to obtain the locations of the other agents (see the setup used in Figure 6.38). The method is similar to position-based visual servoing. The leader computes the team navigation and task coordination. It computes the poses of the followers with respect to the leader, which are then passed to the follower robots that have a controller defining the respective linear and angular velocities to achieve the desired formation/keep the formation.

6.4. Conclusion

In this chapter, we addressed the applications of omnidirectional images to robot localization and navigation. It is clear that, in general, omnidirectional images facilitate localization, detection of obstacles and map building, given the more extensive and complete information about the environment contained in each image. Increased visibility will decrease the instances of occlusion and also increase the range of overlapping visual fields between neighboring images. These advantages can be more specific depending on the system and configuration used to obtain the omnidirectional images. These configurations include systems with multiple cameras and systems with special mirrors and lenses (and one or multiple cameras).

In specific instances, resolution may affect the uncertainty of parameters estimated based on omnidirectional images. A single omnidirectional image of the environment is affected by a smaller angular resolution and that may affect the parameter estimates, especially in outdoor environments where objects are located very far from the imaging system.

Navigation can be significantly improved with the use of this kind of image, and further applications can be expected with the use of deep learning, since the inherent distortions and increased radiometric variability (characteristic of this kind of image) can be more easily dealt with.

6.5. References

Agrawal, A., Taguchi, Y., Ramalingam, S. (2010). Analytical forward projection for axial non-central dioptric and catadioptric cameras. In *Proc. Europ. Conf. Comput. Vis.* Springer, Berlin, Heidelberg, 129–143.

Agrawal, A., Taguchi, Y., Ramalingam, S. (2011). Beyond Alhazens problem: Analytical projection model for non-central catadioptric cameras with quadric mirrors. In *Proc. IEEE Conf. Comput. Vis. Pattern Recognit.*, IEEE, Colorado Springs, CO, USA. 2993–3000.

Baker, S. and Nayar, S. (1998). A theory of catadioptric image formation. In *Proc. IEEE Int. Conf. Computer Vision.* New York, 35–42.

Baker, S. and Nayar, S. (1999). A theory of single-viewpoint catadioptric image formation. *Int. J. Comput. Vision*, 35(2), 175–196.

Baker, P., Fermuller, C., Aloimonos, Y., Pless, R. (2001). A spherical eye from multiple cameras (makes better models of the world). In *Proc. IEEE Conf. Comput. Vis. Pattern Recognit.*, New York.

Betke, M. and Gurvits, L. (1997). Mobile robot localization using landmarks. *IEEE Trans. Robot. Autom.*, 13(2), 251–263.

Bonin-Font, F., Ortiz, A., Oliver, G.O. (2008). Visual navigation for mobile robots: A survey. *J. Intell. Robot. Syst.*, 53(3), 263–296.

Bunschoten, R. and Ben Krose, B. (2003). Visual odometry from an omnidirectional vision system. In *Proc. IEEE Int. Conf. Robot. Automat.*, New York, 1, 577–583.

Calabrese, F. and Indiveri, G. (2005). An omni-vision triangulation-like approach to mobile robot localization. In *Proc. IEEE Int. Symp. Medit. Conf. Contr. Autom. Intell. Contr.*, New York, 604–609.

Cao, Z., Sung, J., Hall, E. (1985). Dynamic omnidirectional vision for mobile robots. In *Intelligent Robots and Computer Vision IV*, Casasent, D.P. (ed.). SPIE, Bellingham, WA.

Caruso, D., Engel, J., Cremers, D. (2015). Large-scale direct slam for omnidirectional cameras. In *Proc. IEEE/RSJ Int. Conf. Intel. Robots Syst.*, 141–148.

Cauchois, C., Brassart, E., Marhic, B., Drocourt, C. (2002). An absolute localization method using a synthetic panoramic image base. In *Proc. IEEE Workshop on Omnidirectional Vision, held in conjunction with ECCV.*, New York, 128–135.

Chahl, J. and Srinivasan, M. (1997). Reflective surfaces for panoramic imaging. *Appl. Opt.*, 36(31), 8275–8285.

Corke, P., Strelow, D., Singh, S. (2004). Omnidirectional visual odometry for a planetary rover. In *Proc. IEEE/RSJ Int. Conf. Intel. Robots Syst.*, New York, 4, 4007–4012.

Courbon, J., Mezouar, Y., Eck, L., Martinet, P. (2008a). Efficient hierarchical localization method in an omnidirectional images memory. In *Proc. IEEE Int. Conf. Robot. Automat.*, New York, 13–18.

Courbon, J., Mezouar, Y., Lequievre, L., Eck, L. (2008b). Navigation of urban vehicle: An efficient visual memory management for large scale environments. In *Proc. IEEE/RSJ Int. Conf. Intel. Robots Syst.*, New York, 1817–1822.

Crombez, N., Caron, G., Mouaddib, E. (2015). Using dense point clouds as environment model for visual localization of mobile robot. In *Proc. 12th Int. Conf. Ubiquitous Robots and Ambient Intelligence.*, New York, 40–45.

Das, A.K., Fierro, R., Kumar, V., Ostrowski, J.P., Spletzer, J., Taylor, C.J. (2002). A vision-based formation control framework. *IEEE Trans. Robot. Autom.*, 18(5), 813–825.

Engel, J., Schops, T., Cremers, D. (2014). LSD-SSLAM: Large-scale direct monocular SLAM. In *Proc. Eur. Conf. Comput. Vis.* Springer, Berlin, Heidelberg, 834–849.

Fasogbon, P. and Aksu, E. (2019). Calibration of fisheye camera using entrance pupil. In *Proc. IEEE Int. Conf. Image Proc.*, New York, 469–473.

Fernandez-Labrador, C., Facil, J.M., Perez-Yus, A., Demonceaux, C., Civera, J., Guerrero, J.J. (2020). Corners for layout: End-to-end layout recovery from 360 images. *IEEE Rob. Autom. Lett.*, 5(2), 1255–1262.

Fitzgibbon, A.W. (2001). Simultaneous linear estimation of multiple view geometry and lens distortion. In *Proc. IEEE Conf. Comput. Vis. Pattern Recognit.*, New York, 1, 125–132.

Gamallo, C., Mucientes, M., Regueiro, C.V. (2015). Omnidirectional visual SLAM under severe occlusions. *Robot. Autonom. Syst.*, 65, 76–87.

Gaspar, J., Decco, C., Okamoto, J., Santos-Victor, J. (2002). Constant resolution omnidirectional cameras. In *Proc. IEEE Workshop on Omnidirectional Vision*, New York, 27–34.

Gava, C.C., Vassallo, R.F., Roberti, F., Carelli, R., Bastos-Filho, T.F. (2007). Nonlinear control techniques and omnidirectional vision for team formation on cooperative robotics. In *Proc. IEEE Int. Conf. Robot. Automat.*, New York, 2409–2414.

Gennery, D.B. (2006). Generalized camera calibration including fish-eye lenses. *Int. J. Comput. Vision*, 68(3), 239–266.

Geyer, C. and Daniilidis, K. (2000). A unifying theory for central panoramic systems and practical implications. *Proc. 6th Eur. Conf. Comput. Vis.*, Dublin.

Gonçalves, N. (2010). On the reflection point where light reflects to a known destination on quadratic surfaces. *Opt. Lett.*, 35(2), 100–102.

Grossberg, M.D. and Nayar, S.K. (2001). A general imaging model and a method for finding its parameters. In *Proc. IEEE Int. Conf. Comp. Vis.*, 2, 108–115.

Guerbas, S.-E., Crombez, N., Caron, G., Mustapha Mouaddib, E. (2021). Photometric Gaussian mixtures for direct virtual visual servoing of omnidirectional camera. In *IEEE Conf. Comp. Vis. Patt. Recogn. Workshops*, New York.

Heng, L., Lee, G., Pollefeys, M. (2014). Self-calibration and visual SLAM with a multi-camera system on a micro aerial vehicle. In *Proc. Robotics: Science and Systems*. Berkeley, CA [Online]. Available at: http://roboticsproceedings.org/.

Heng, L., Lee, G., Pollefeys, M. (2015). Self-calibration and visual SLAM with a multi-camera system on a micro aerial vehicle. *Auton. Robot.*, 39, 259–277.

Hicks, R. (2005). Designing a mirror to realize a given projection. *J. Opt. Soc. Am. A.*, 22(2), 323–330.

Hicks, R. and Bajcsy, R. (1999). Reflective surfaces as computational sensors. In *Workshop on Perception for Mobile Agents*, 82–86.

Hicks, R. and Bajcsy, R. (2000). Catadioptric sensors that approximate wide-angle perspective projections. In *Proc. IEEE Workshop on Omnidirectional Vision*, New York, 97–103.

Hicks, R. and Coletta, M. (2013). Computational photography with panoramic sensors that have uniform resolution with respect to unwarping transformations. *J. Math. Imaging Vision*, 46, 121–127.

Hicks, R. and Perline, R. (2001). Geometric distributions for catadioptric sensor design. In *Proc. IEEE Conf. Comput. Vis. Pattern Recognit.*, New York, 1.

Hicks, R. and Perline, R. (2004). The method of vector fields for catadioptric sensor design with applications to panoramic imaging. In *Proc. IEEE Conf. Comput. Vis. Pattern Recognit.*, New York, 2.

Hicks, R., Perline, R.K., Coletta, M.L. (2001). Catadioptric sensors for panoramic viewing. In *Computing and Information Technologies*, Antoniou, G. and Deremer, D. (eds). World Scientific, Montclair State University, NJ.

Hong, J., Tan, X., Pinette, B., Weiss, R., Riseman, E. (1991). Image-based homing. In *Proc. IEEE Int. Conf. Robot. Automat.*, 1, 620–625.

Hongdong, L., Hartley, R., Kim, J.-H. (2008). A linear approach to motion estimation using generalized camera models. In *Proc. IEEE Conf. Comp. Vis. Pattern Recogn.*, 1–8.

Hughes, C., Denny, P., Jones, E., Glavin, M. (2010). Accuracy of fish-eye lens models. *Appl. Opt.*, 49(17), 3338–3347.

Jang, G., Kim, S., Kim, J., Kweon, I.-S. (2005). Metric localization using a single artificial landmark for indoor mobile robots. In *Proc. IEEE/RSJ Int. Conf. Intel. Robots Syst.*, Edmonton, 2857–2862.

Jayasuriya, M., Ranasinghe, R., Dissanayake, G. (2020). Active perception for outdoor localisation with an omnidirectional camera. In *Proc. IEEE/RSJ Int. Conf. Intel. Robots Syst.*, New York, 4567–4574.

Kannala, J. and Brandt, S. (2006). A generic camera model and calibration method for conventional, wide-angle, and fish-eye lenses. *IEEE Trans. Pattern Anal. Mach. Intell.*, 28(8), 1335–1340.

Kazik, T., Kneip, L., Nikolic, J., Pollefeys, M., Siegwart, R. (n.d.). Real-time 6D stereo visual odometry with non-overlapping fields of view. In *Proc. IEEE Conf. Comput. Vis. Pattern Recognit.*, New York, 1529–1536.

Klingner, B., Martin, D., Roseborough, J. (2013). Street view motion-from-structure-from-motion. In *Proc. IEEE Int. Conf. Computer Vision*, New York, 953–960.

de La Garanderie, G., Abarghouei, A., Breckon, T. (2018). Eliminating the blind spot: Adapting 3D object detection and monocular depth estimation to 360° panoramic imagery. In *Proc. Eur. Conf. Comput. Vis.*, 812–830.

Laina, I., Rupprecht, C., Belagiannis, V., Tombari, F., Navab, N. (2016). Deeper depth prediction with fully convolutional residual networks. In *Proc. Int. Conf. 3D Vision*. IEEE, New York, 239–248.

Lee, G.H., Faundorfer, F., Pollefeys, M. (2013). Motion estimation for self-driving cars with a generalized camera. In *Proc. IEEE Conf. Comput. Vis. Pattern Recognit.*, New York, 2746–2753.

Lee, Y., Jeong, J., Yun, J., Cho, W., Yoon, K.-J. (2019). SpherePHD: Applying CNNs on a spherical polyhedron representation of 360-degree images. In *Proc. IEEE Conf. Comput. Vis. Pattern Recognit.*, 9173–9181.

Li, S. and Isago, T. (2007). Qualitative localization by full-view spherical image. In *Proc. IEEE Int. Conf. Autom. Sc. Eng.*, New York, 566–571.

Li, M., Imou, K., Wakabayashi, K., Yokoyama, S. (2010). A new agricultural vehicle localization system. In *Proc. World Automation Congress.* IEEE, New York, 333–338.

Loevsky, I. and Shimshoni, I. (2010). Reliable and efficient landmark-based localization for mobile robots. *Robot. Autonom. Syst.*, 58(5), 520–528.

Lukierski, R., Leutenegger, S., Davison, A. (2015). Rapid free-space mapping from a single omnidirectional camera. In *Proc. Eur. Conf. Mob. Rob.*, IEEE, New York, 1–8.

Marhic, B., Mouaddib, E., Pégard, C. (1998). A localisation method with an omnidirectional vision sensor using projective invariant. In *Proc. IEEE/RSJ Int. Conf. Intel. Robots Syst.*, New York, 2, 1078–1083.

Mariottini, G.L., Morbidi, F., Prattichizzo, D., Pappas, G.J., Daniilidis, K. (2007). Leader-follower formations: Uncalibrated vision-based localization and control. In *Proc. IEEE Int. Conf. Robot. Automat.*, New York, 2403–2408.

Mariottini, G.L., Morbidi, F., Prattichizzo, D., Vander Valk, N., Michael, N., Pappas, G., Daniilidis, K. (2009). Vision-based localization for leader-follower formation control. *IEEE Trans. Robotics*, 25(6), 1431–1438.

Marques, C.F. and Lima, P.U. (2001). A localization method for a soccer robot using a vision-based omni-directional sensor. In *RoboCup 2000: Robot Soccer World Cup IV*, Stone, P., Balch, T., Kraetzschmar, G. (eds). Springer, Melbourne.

Matsuki, H., von Stumberg, L., Usenko, V., Stuckler, J., Cremers, D. (2018). Omnidirectional DSO: Direct sparse odometry with fisheye cameras. *IEEE Rob. Autom. Lett..* 3(4), 3693–3700.

Meilland, M., Comport, A., Rives, P. (2015). Dense omnidirectional RGB-D mapping of large-scale outdoor environments for real-time localization and autonomous navigation. *J. Field Rob.*, 32(4), 474–503.

Menegatti, E., Pretto, A., Pagello, E. (2004). Testing omnidirectional vision-based Monte Carlo localization under occlusion. In *Proc. IEEE/RSJ Int. Conf. Intel. Robots Syst.*, New York, 3, 2487–2493.

Menegatti, E., Pretto, A., Scarpa, A., Pagello, E. (2006). Omnidirectional vision scan matching for robot localization in dynamic environments. *IEEE Trans. Robotics*, 22(3), 523–535.

Micusik, B. and Pajdla, T. (2003). Estimation of omnidirectional camera model from epipolar geometry. In *Proc. IEEE Conf. Comput. Vis. Pattern Recognit.*, 1.

Micusik, B. and Pajdla, T. (2004). Autocalibration & 3D reconstruction with non-central catadioptric cameras. In *Proc. IEEE Conf. Comput. Vis. Pattern Recognit.*, New York, 1, 58–65.

Milford, M. (2008). *Robot Navigation from Nature-Simultaneous Localisation, Mapping, and Path Planning Based on Hippocampal Models*. Springer, Berlin, Heidelberg.

Miraldo, P. and Araújo, H. (2014). Planar pose estimation for general cameras using known 3D lines. In *Proc. IEEE/RSJ Int. Conf. Intel. Robots Syst.*, New York, 4234–4240.

Miraldo, P. and Araújo, H. (2015). Generalized essential matrix: Properties of the singular value decomposition. *Image Vision Comput.*, 34, 45–50.

Miraldo, P. and Cardoso, J.R. (2020). On the generalized essential matrix correction: An efficient solution to the problem and its applications. *J. Math. Imaging Vision*, 62, 1107–1120.

Miraldo, P., Araújo, H., Queiro, J. (2011). Point-based calibration using a parametric representation of the general imaging model. In *Proc. IEEE Int. Conf. Computer Vision*, 2304–2311.

Miraldo, P., Araújo, H., Gonçalves, N. (2015). Pose estimation for general cameras using lines. *IEEE Trans. Cybern.*, 45(10), 2156–2164.

Miraldo, P., Ramalingam, S., Eiras, F. (2018). Analytical modeling of vanishing points and curves in catadioptric cameras. In *Proc. IEEE Conf. Comput. Vis. Pattern Recognit.*, New York, 2012–2021.

Miyamoto, K. (1964). Fish eye lens. *J. Opt. Soc. Am.*, 54(8), 1060–1061.

Mouragnon, E., Lhuillier, M., Dhome, M., Dekeyser, F., Sayd, P. (2009). Generic and real-time structure from motion using local bundle adjustment. *Image Vision Comput.*, 27(8), 1178–1193.

Munguía, R., López-Franco, C., Nuño, E., Lopéz-Franco, A. (2017). Method for SLAM based on omnidirectional vision: A delayed-EKF approach. *J. Sens.* [Online]. Available at: www.hindawi.com/journals/js/2017/7342931/.

Nalwa, V. (1996). A true omnidirectional viewer. *AT&T Bell Laboratories, Tech. Rep. BL0115500-960115-01*, January.

Nayar, S. (1997). Catadioptric omnidirectional camera. In *Proc. IEEE Conf. Comput. Vis. Pattern Recognit.*, 482–488.

Ollis, M., Herman, H., Singh, S. (1999). Analysis and design of panoramic stereo vision using equi-angular pixel cameras. Technical Report CMU-RI-TR-99-04, Carnegie Mellon University, Pittsburgh, PA.

Paya, L., Gil, A., Reinoso, O. (2017a). A state-of-the-art review on mapping and localization of mobile robots using omnidirectional vision sensors. *J. Sens.*, 1–20.

Paya, L., Reinoso, O., Jimenez, L., Julia, M. (2017b). Estimating the position and orientation of a mobile robot with respect to a trajectory using omnidirectional imaging and global appearance. *PLoS ONE*, 12(5), 1–25.

Pless, R. (2003). Using many cameras as one. In *Proc. IEEE Conf. Comput. Vis. Pattern Recognit.*, New York, 2.

Ramalingam, S., Bouaziz, S., Sturm, P., Brand, M. (2010). SKYLINE2GPS: Localization in urban canyons using omni-skylines. In *Proc. IEEE/RSJ Int. Conf. Intel. Robots Syst.*, New York, 3816–3823.

Ramezani, M., Khoshelham, K., Kneip, L. (2017). Omnidirectional visual-inertial odometry using multi-state constraint Kalman filter. In *Proc. IEEE/RSJ Int. Conf. Intel. Robots Syst.*, 1317–1323.

Rituerto, A., Puig, L., Guerrero, J.J. (2010a). Visual SLAM with an omnidirectional camera. In *Proc. IEEE Int. Conf. Pattern Recognit.*, New York, 348–351.

Rituerto, A., Puig, L., Guerrero, J.J. (2010b). Comparison of omnidirectional and conventional monocular systems for visual SLAM. In *10th Workshop on Omnidirectional Vision, Camera Networks and Non-classical Cameras (OMNIVIS)*, January.

Scaramuzza, D. and Siegwart, R. (2008). Appearance-guided monocular omnidirectional visual odometry for outdoor ground vehicles. *IEEE Trans. Robotics*, 24(5), 1015–1026.

Scaramuzza, D., Fraundorfer, F., Siegwart, R. (2009). Real-time monocular visual odometry for on-road vehicles with 1-point RANSAC. In *Proc. IEEE Int. Conf. Robot. Automat.*, Zaragoza, 4293–4299.

Seok, H. and Lim, J. (2019). ROVO: Robust omnidirectional visual odometry for wide-baseline wide-FOV camera systems. In *Proc. IEEE Int. Conf. Robot. Automat.*, New York, 6344–6350.

Siegwart, R., Nourbakhsh, I., Scaramuzza, D. (2011). *Introduction to Autonomous Mobile Robots*, 2nd edition. MIT Press, Cambridge, MA.

Strelow, D., Mishler, J., Singh, S., Herman, H. (2001). Extending shape-from-motion to noncentral onmidirectional cameras. In *Proc. IEEE/RSJ Int. Conf. Intel. Robots Syst.*, New York, 4, 2086–2092.

Sturm, P. and Ramalingam, S. (2004). A generic concept for camera calibration. In *Proc. Eur. Conf. Comput. Vis.* Springer, Berlin, Heidelberg, 1–13.

Sun, C., Hsiao, C.-W., Sun, M., Chen, H.-T. (2019). HorizonNet: Learning room layout with 1D representation and pano stretch data augmentation. In *Proc. IEEE Conf. Comput. Vis. Pattern Recognit.*, New York, 1047–1056.

Swaninathan, R., Grossberg, M.D., Nayar, S.K. (2003). A perspective on distortions. In *Proc. IEEE Conf. Comput. Vis. Pattern Recognit.*, New York, 2, 594–601.

Swaminathan, R., Grossberg, M.D., Nayar, S.K. (2006). Non-single viewpoint catadioptric cameras: Geometry and analysis. *Int. J. Comput. Vision*, 66(3), 211–229.

Tardif, J.-P., Pavlidis, Y., Daniilidis, K. (2008). Monocular visual odometry in urban environments using an omnidirectional camera. In *Proc. IEEE/RSJ Int. Conf. Intel. Robots Syst.*, 2531–2538.

Tateno, K., Navab, N., Tombari, F. (2018). Distortion-aware convolutional filters for dense prediction in panoramic images. In *Proc. Eur. Conf. Comput. Vis.*, 732–750.

Thirthala, S. and Pollefeys, M. (2012). Radial multi-focal tensors. *Int. J. Comput. Vision*, 96(2), 195–211.

Usenko, V., Demmel, N., Cremers, D. (2016). The double sphere camera model. In *Proc. IEEE Int. Conf. 3D Vision*, 552–560.

Vidal, R., Shakernia, O., Sastry, S. (2003). Formation control of nonholonomic mobile robots with omnidirectional visual servoing and motion segmentation. In *Proc. IEEE Int. Conf. Robot. Automat.*, New York, 1, 584–589.

Vidal, R., Shakernia, O., Sastry, S. (2004). Following the flock [formation control]. *IEEE Rob. Autom. Mag.*, 11(4), 14–20.

Wang, S., Clark, R., Wen, H., Trigoni, N. (2018). End-to-end, sequence-to-sequence probabilistic visual odometry through deep neural networks. *Int. J. Robot. Res.*, 37(4–5), 513–542.

Winters, N., Gaspar, J., Lacey, G., Santos-Victor, J. (2000). Omni-directional vision for robot navigation. In *Proc. IEEE Workshop on Omnidirectional Vision (OMNIVIS)*, New York, 21–28.

Won, C., Ryu, J., Lim, J. (2019). OmniMVS: End-to-end learning for omnidirectional stereo matching. In *Proc. IEEE Int. Conf. Computer Vision*, New York, 8986–8995.

Won, C., Ryu, J., Lim, J. (2020a). End-to-end learning for omnidirectional stereo matching with uncertainty prior. *IEEE Trans. Pattern Anal. Mach. Intell.*, 43(11), 3850–3862.

Won, C., Seok, H., Cui, Z., Pollefeys, M., Lim, J. (2020b). OmniSLAM: Omnidirectional localization and dense mapping for wide-baseline multi-camera systems. In *Proc. IEEE Int. Conf. Robot. Automat.*, 559–566.

Wu, C.-J. and Tsai, W.-H. (2009). Location estimation for indoor autonomous vehicle navigation by omni-directional vision using circular landmarks on ceilings. *Robot. Autonom. Syst.*, 57(5), 546–555.

Wu, C.-J. and Tsai, W.-H. (2010). An omni-vision based localization method for automatic helicopter landing assistance on standard helipads. In *Proc. 2nd Int. Conf. Comp. Autom. Eng.*, New York, 3, 327–332.

Xiang, Z., Dai, X., Gong, X. (2013). Noncentral catadioptric camera calibration using a generalized unified model. *Opt. Lett.*, 38(9), 1367–1369.

Yagi, Y. and Kawato, S. (1990). Panorama scene analysis with conic projection. In *Proc. IEEE/RSJ Int. Conf. Intel. Robots Syst.*, New York, 1, 181–187.

Yagi, Y. and Yachida, M. (1991). Real-time generation of environmental map and obstacle avoidance using omnidirectional image sensor with conic mirror. In *Proc. IEEE Conf. Comput. Vis. Pattern Recognit.*, New York, 160–165.

Yagi, Y., Kawato, S., Tsuji, S. (1991a). Collision avoidance using omnidirectional image sensor (COPIS). In *Proc. IEEE Int. Conf. Robot. Automat.*, New York, 1, 910–915.

Yagi, Y., Nishizawa, Y., Yachida, M. (1991b). Estimating location and avoiding collision against unknown obstacles for the mobile robot using omnidirectional image sensor COPIS. In *Proc. IEEE/RSJ Int. Conf. Intel. Robots Syst.*, New York, 2, 909–914.

Yagi, Y., Nishizawa, Y., Yachida, M. (1991c). Estimation of free space for the mobile robot using omnidirectional image sensor COPIS. In *Proc. Int. Conf. Ind. Elect. Contr. Instrum.*, New York, 2, 1329–1334.

Yagi, Y., Kawato, S., Tsuji, S. (1994). Real-time omnidirectional image sensor (COPIS) for vision-guided navigation. *IEEE Trans. Robot. Autom.*, 10(1), 11–22.

Yagi, Y., Nishizawa, Y., Yachida, M. (1995). Map-based navigation for a mobile robot with omnidirectional image sensor COPIS. *IEEE Trans. Robot. Autom.*, 11(5), 634–648.

Ying, X. and Hu, Z. (2004). Can we consider central catadioptric cameras and fisheye cameras within a unified imaging model. In *Proc. Eur. Conf. Comp. Vis.*, 442–455.

Zhang, Z., Rebecq, H., Forster, C., Scaramuzza, D. (2016). Benefit of large field-of-view cameras for visual odometry. In *Proc. IEEE Int. Conf. Robot. Automat.*, New York, 801–808.

Zheng, Y. and Tsuji, S. (1992). Panoramic representation for route recognition by a mobile robot. *Int. J. Comput. Vision*, 9(1), 55–76.

Zioulis, N., Karakottas, A., Zarpalas, D., Daras, P. (2018). OmniDepth: Dense depth estimation for indoors spherical panoramas. In *Proc. Eur. Conf. Comp. Vis.* Springer, Berlin, Heidelberg, 453–471.

Zou, C., Su, J.-W., Peng, C.-H., Colburn, A., Shan, Q., Wonka, P., Chu, H.-K., Hoiem, D. (2021). Manhattan room layout reconstruction from a single 360 image: A comparative study of state-of-the-art methods. *Int. J. Comput. Vision*, 129(5), 1410–1431.

Conclusion and Perspectives

Fabio MORBIDI and Pascal VASSEUR

MIS Laboratory, University of Picardie Jules Verne, Amiens, France

> *The Kuang program spurted from tarnished cloud,*
> *Case's consciousness divided like beads of mercury*
> *arcing above an endless beach the color of the dark*
> *silver clouds. His vision was spherical, as though a single*
> *retina lined the inner surface of a globe that contained all things,*
> *if all things could be counted.*
> Neuromancer, William GIBSON, 1984, Chapter 23

C.1. Epilogue

This book presents, in an accessible but rigorous form, the theoretical foundations and the classical applications of omnidirectional vision. Its main objective is to fill the gap between theory and practice, left by previous books and monographs, and to present an up-to-date overview of the state of the art, which reflects current knowledge. It is our hope that this book will contribute to the education of future practitioners and will inspire a new generation of researchers and technology leaders in computer vision and robotics in the years to come.

Omnidirectional Vision,
coordinated by Pascal VASSEUR and Fabio MORBIDI. © ISTE Ltd 2023.

C.2. Prospects and challenges ahead

Although research progressed at a slower pace in the 2010s, new applications and important advances in computer science and artificial intelligence have sparked renewed interest in the field of omnidirectional vision in the early 2020s. In fact, driven by the resounding success of deep learning in computer vision (and more specifically convolutional neural networks (CNNs) (Li et al. 2022)), *machine learning techniques* have also made their way into 360° vision. For example, 3D scene geometry understanding from a single 360° image (da Silveira et al. 2023) is the subject of active research (Sun et al. 2019; Fernandez-Labrador 2020; Zou et al. 2021; Tran 2021; Rey-Area et al. 2022; Shen et al. 2022), with attractive applications in many areas (autonomous driving, real estate advertising and military intelligence).

The rise of graph-structured data, and the emergence of graph signal processing (Cheung et al. 2018), which aims to develop tools for processing data defined on irregular graph domains, extending techniques (transforms, sampling, filtering, etc.) used for conventional signals, have also had strong repercussions in image processing (Ortega et al. 2018) and learning systems (Wu et al. 2021) (graph neural networks). Omnidirectional vision requires the processing of omnidirectional signals lying on a sphere, and it is expected to greatly benefit from these new tools. Graph-based representations have been used for the classification of omnidirectional images (Khasanova and Frossard 2017). Moreover, toward a generalization of deep learning models to non-Euclidean domains, CNNs have been recently adapted to spherical signals (Khasanova and Frossard 2019; Eder and Frahm 2019; Gerken et al. 2022) for 360° image compression (Bidgoli et al. 2022), handwritten-digit and sport classification, saliency detection, and gaze tracking in 360° videos (Xu et al. 2022).

Finally, optics enabling 360° vision are still waiting to be fully exploited in non-conventional cameras (e.g. event-based, polarization, and multispectral cameras).

These are interesting avenues for future research, with far-reaching ramifications in engineering and natural sciences. This is an exciting time for the field of omnidirectional vision and a bright future lies ahead.

C.3. References

Bidgoli, N., Azevedo, R., Maugey, T., Roumy, A., Frossard, P. (2022). OSLO: On-the-sphere learning for omnidirectional images and its application to 360° image compression. *IEEE Trans. Image Process.*, 31, 5813–5827.

Cheung, G., Magli, E., Tanaka, Y., Ng, M. (2018). Graph spectral image processing. *Proc. IEEE*, 106(5), 907–930.

Eder, M. and Frahm, J.-M. (2019). Convolutions on spherical images. In *Proc. IEEE Conf. Comput. Vis. Pattern Recognit. Workshops*. IEEE, New York.

Fernandez-Labrador, C., Facil, J.M., Perez-Yus, A., Demonceaux, C., Civera, J., Guerrero, J.J. (2020). Corners for layout: End-to-end layout recovery from 360 images. *IEEE Rob. Autom. Lett.*, 5(2), 1255–1262.

Gerken, J., Carlsson, O., Linander, H., Ohlsson, F., Petersson, C., Persson, D. (2022). Equivariance versus augmentation for spherical images. In *Proc. 39th Int. Conf. Mach. Learn.*, 162, 7404–7421. Baltimore, MA.

Khasanova, R. and Frossard, P. (2017). Graph-based classification of omnidirectional images. In *Proc. IEEE Int. Conf. Comp. Vis.*, 869–878. IEEE, New York.

Khasanova, R. and Frossard, P. (2019). Geometry aware convolutional filters for omnidirectional images representation. In *Proc. Mach. Learn. Res.*, 97, 3351–3359. Long Beach, CA.

Li, Z., Liu, F., Yang, W., Peng, S., Zhou, J. (2022). A survey of convolutional neural networks: Analysis, applications, and prospects. *IEEE Trans. Neural Networks Learn. Syst.*, 33(12), 6999–7019.

Ortega, A., Frossard, P., Kovačević, J., Moura, J., Vandergheynst, P. (2018). Graph signal processing: Overview, challenges, and applications. *Proc. IEEE*, 106(5), 808–828.

Rey-Area, M., Yuan, M., Richardt, C. (2022). 360MonoDepth: High-resolution 360° monocular depth estimation. In *Proc. IEEE Conf. Comput. Vis. Pattern Recognit.*, 3762–3772. IEEE, New York.

Shen, Z., Lin, C., Nie, L., Liao, K., Zhao, Y. (2022). Neural contourlet network for monocular 360° depth estimation. *IEEE Trans. Circuits Syst. Video Technol.*, 32(12), 8574–8585.

da Silveira, T., Pinto, P., Murrugarra-Llerena, J., Jung, C. (2023). 3D scene geometry estimation from 360° imagery: A survey. *ACM Comput. Surv.*, 55(4), 1–39.

Sun, C., Hsiao, C.-W., Sun, M., Chen, H.-T. (2019). HorizonNet: Learning room layout with 1D representation and pano stretch data augmentation. In *Proc. IEEE Conf. Comput. Vis. Pattern Recognit.*, 1047–1056. IEEE, New York.

Tran, P. (2021). SSLayout360: Semi-supervised indoor layout estimation from 360° panorama. In *Proc. IEEE Conf. Comput. Vis. Pattern Recognit.*, 15353–15362. IEEE, New York.

Wu, Z., Pan, S., Chen, F., Long, G., Zhang, C., Philip, S. (2021). A comprehensive survey on graph neural networks. *IEEE Trans. Neural Networks Learn. Syst.*, 32(1), 4–24.

Xu, Y., Zhang, Z., Gao, S. (2022). Spherical DNNs and their applications in 360° images and videos. *IEEE Trans. Pattern Anal. Mach. Intell.*, 44(10), 7235–7252.

Zou, C., Su, J.-W., Peng, C.-H., Colburn, A., Shan, Q., Wonka, P., Chu, H.-K., Hoiem, D. (2021). Manhattan room layout reconstruction from a single 360 image: A comparative study of state-of-the-art methods. *Int. J. Comput. Vision*, 129(5), 1410–1431.

List of Authors

Helder Jesus ARAÚJO
Institute of Systems and Robotics
University of Coimbra
Portugal

Fatima AZIZ
XLIM Laboratory
University of Limoges
France

Guillaume CARON
MIS Laboratory
University of Picardie Jules Verne
Amiens
France
and
CNRS-AIST Joint Robotics
Laboratory (JRL)
Tsukuba
Japan

Nathan CROMBEZ
CIAD Laboratory
University of Technology
of Belfort-Montbéliard
France

Cédric DEMONCEAUX
ImViA Laboratory
University of Burgundy
Dijon
France

Sio-hoi IENG
Institut de la Vision
University Pierre and Marie Curie
Paris
France

Ouiddad LABBANI-IGBIDA
XLIM Laboratory
University of Limoges
France

Maxime LHUILLIER
University Clermont Auvergne
CNRS
Institut Pascal
Clermont-Ferrand
France

Pedro MIRALDO
Mitsubishi Electric Research
Laboratories (MERL)
Boston
USA

Fabio MORBIDI
MIS Laboratory
University of Picardie Jules Verne
Amiens
France

Peter STURM
Inria Grenoble Rhône-Alpes
Montbonnot-Saint-Martin
France

Pascal VASSEUR
MIS Laboratory
University of Picardie Jules Verne
Amiens
France

Index

Printed and bound by CPI Group (UK) Ltd, Croydon, CR0 4YY

20/12/2023

08212804-0002